Blazing Heritage

Blazing Heritage

A History of Wildland Fire in the National Parks

HAL K. ROTHMAN

2007

OXFORD
UNIVERSITY PRESS

Oxford University Press, Inc., publishes works that further
Oxford University's objective of excellence
in research, scholarship, and education.

Oxford New York
Auckland Cape Town Dar es Salaam Hong Kong Karachi
Kuala Lumpur Madrid Melbourne Mexico City Nairobi
New Delhi Shanghai Taipei Toronto

With offices in
Argentina Austria Brazil Chile Czech Republic France Greece
Guatemala Hungary Italy Japan Poland Portugal Singapore
South Korea Switzerland Thailand Turkey Ukraine Vietnam

Published by Oxford University Press, Inc.
198 Madison Avenue, New York, New York 10016

www.oup.com

Library of Congress Cataloging-in-Publication Data
Rothman, Hal, 1958–
Blazing heritage : a history of wildland fire in the national parks /
Hal K. Rothman.
p. cm.
Includes bibliographical references and index.
ISBN 978-0-19-531116-7
1. Wildfires—United States—History. 2. Forest fires—United States—
History. 3. National parks and reserves—United States—History.
4. Fire management—United States—History. I. Title.
SD421.3.R68 2007
363.37'90973—dc22 2006050728

2 4 6 8 9 7 5 3 1

Printed in the United States of America
on acid-free paper

For Lauralee, Talia, and Brent,
whose strength and courage exceed anything I have ever seen

Acknowledgments

This project would never have happened without the insight and support of the National Park Service (NPS) and its officials at the National Interagency Fire Center. No one was more important to this project than Rick Gale, who made it a personal passion as he prepared for his retirement. With the support of the chief historian, Dwight Pitcaithley, who saw the value in the project and envisioned what it could become, the project got off the ground. NPS bureau historian Janet McDonnell played the most instrumental role, shaping the project and handling the administrative details as well as collating the comments of countless reviewers. Steve Pyne served as senior consultant on this project, providing me with the benefit of his unparalleled knowledge about fire. Steve and I argued almost endlessly about what belonged in here; in the end, I deferred to his vastly superior knowledge more often than I asserted the author's prerogative to tell it my way. Roberta D'Amico of the National Interagency Fire Center picked up where Rick Gale left off, providing access to people and information in all kinds of ways. I was fortunate to be able to interview a number of important fire luminaries—Norm Christensen of Duke University, Jim Agee of the University of Washington, former NPS officials Bruce Kilgore and Bob Barbee—as well as countless other people who generously gave their time to this project, including Jan van Wagtendonk, John Lissoway, Craig Allen, Brad Cella, James Cook, Bob Linn, and Kevin McKibben. Without their insights and help and that of so many others, this project would be much less than it is.

This was also a teaching project, and I was fortunate to have the benefit of wonderful assistants. Lincoln Bramwell, Brenna Lissoway, and Lesley Argo did the bulk of the research on the project. They not only accompanied me to the National Archives in Washington, D.C., they spread out over the nation to engage in research in parks and repositories. Even after their time on the project was finished, they continued to offer ideas and insights and especially to correct any errors I made. I hope this experience was as rewarding for them as it was for me. Two others served as project administrators. Dave Sproul handled the details during the research phase, dealing with university travel people and airlines, car rentals, hotels, and motels—all with his usual verve. Mike Childers came in during the last two years of the project and proved to be an invaluable organizer and administrator. Both of them freed me to do the writing on the project.

During the last stages of this project, I was diagnosed with ALS, Lou Gehrig's disease. This has taken me from complete independence to nearly complete dependence in the space of one year. As a result, nothing I do is by myself, alone, if it ever was. I rely on an entire staff of people—family, friends, and students—not only to get me through the day, but also to let me do my work. No one has done more on this project than "my hands," Leisl Carr, who has faithfully translated the increasing gibberish that comes out of my mouth into the words I want on the page. My dear friend Nancy Jackson edited the manuscript and helped me to shape it prior to publication. She has edited nearly a dozen of my works now, and even though she has left academic publishing, her influence on me and so many others remains strong. Susan Ferber at Oxford University Press provided an excellent read and set of comments on the manuscript that vastly improved it.

At UNLV, so many others have helped in their way: President Carol Harter, Dean Ed Shoben, and my chair, Gene Moehring, have helped to clear the way for an increasingly physically handicapped person to operate as a fully functioning part of the university. Dave Sproul has taken a phenomenal amount of his personal time to help me get through the day, taking the term Ph.D. and turning it into "Personal Helper Dude." Kathy Adkins, Lynette Webber, and Bobbie Betts have helped me with daily details like filing grades and opening the mail. Andy Fry has been a consistent source of optimism and good cheer. Aaron McArthur, Charlie Deitrich, Billy Brost, and Mike Childers all gave something of themselves to let me get a little more done. So many others, colleagues and friends, have made my continued presence possible. I hope this will not be the last book I write.

Finally, there is my family. My wife, Lauralee, has born the brunt of this beast of a disease, and my love and gratitude to her know no bounds. Our children, Talia and Brent, have had to grow up well before their time, and they have done me the honor of continuing to live their lives in the face of unparalleled adversity. They too have my undying affection and love, as well as my deepest gratitude. I could not continue every day without them.

Contents

Blazing Heritage

Introduction

The National Parks and Fire

On July 31, 1926, with a high wind blowing, the gasoline tank on a privately owned logging truck in Glacier National Park exploded, igniting a fire that quickly spread. It crossed Fish Creek and reached Lake McDonald, one of the primary features of the park. Horace M. Albright, associate director of the National Park Service, superintendent of Yellowstone National Park, and the driving force for the fledgling agency, feared that the fire would ruin that scenic section of Glacier, although the terrain it contained was "the most difficult [he] had ever seen" for fighting fire.[1]

This particular fire devastated the land it burned. It was "intensely hot," Albright noted. The understory, the combustible material that builds up on the ground in the absence of sporadic fire, was particularly thick, mute testimony to the success of localized fire suppression. "The timber is thick and heavy, and the ground is covered with brush, down timber, and deep humus of pine needles and rotten leaves," Albright observed.[2]

Albright recognized the danger of the Lake McDonald fire and mustered all the resources that he could find. He drafted one hundred men from construction crews, moving them to the west side of Lake McDonald on August 5. In the next twenty-four hours, they trenched the fire to the summit of Howe Ridge, blunting its advance. At the same time, high winds—the most powerful that many long-time Montanans recalled ever experiencing—drove another fire, called the West Huckleberry fire, down the north slope of Apgar Mountain until it merged with another branch of the Lake McDonald fire. The meeting of the two fires, Albright told NPS director Stephen T. Mather, "compelled considerable readjustment of our fire fighting organization." Digging trenches in front of these fires required the redeployment of a number of the men and took almost four days to accomplish. By August 14, Albright reported with relief that he had the majority of the fires under control. Only a little trenching work remained, he reported, and nearly all the fires were contained. Rain eventually accomplished the rest.[3]

Thirty-nine years later, Yellowstone National Park began to experiment with "controlled burning," the previously heretical idea that people might intentionally set fires to manage park ecology. Yellowstone undertook an aspen restoration program that relied on intentionally set fire. The program had two objectives: to ascertain if the burning of a former stand of aspen, taken over by conifers, would cause the area to revert to aspen and to see if burning would enhance the "number and vigor of Aspen sprouts in a decadent stand" as park biologist Robert E. Howe described it. The park planned a large burn, but the summer of 1965 remained too wet to sustain sufficient fire. Scaling down his expectations, Howe carefully selected a location, a five-acre patch on a north-facing slope just south of the old Tower Falls road. Conifers had taken over a stand of mature aspen. Surrounded by grasslands, it provided an excellent opportunity for an experiment in controlled burning.[4]

Finding optimal conditions proved to be difficult. As September ended, the chance to burn seemed to have passed. In early October, Howe tried to set up the burn, only to be thwarted by a six-inch snowfall. Undaunted, the biologist kept trying. At 1:45 p.m. on October 28, in gusting winds of up to eighteen miles per hour and with a relative humidity of 50 percent, a crew began to intentionally burn timber inside Yellowstone National Park. The men sprayed the conifers with diesel oil and ignited them. The fire went up into the crowns and burned about ten trees closely clumped together. As the gusty winds died down, the fire followed, and despite efforts to reignite it, the crew recorded little success. After three hours and 200 gallons of diesel fuel, they decided that the conditions "weren't going to be favorable this year for a burn," Howe told his superiors.[5]

Much more than thirty-nine years separated these two vignettes. A revolution in thinking had also taken place. In 1926, the idea of igniting fire in a national park was anathema. Fire was the enemy; national park officials from the top to the bottom regarded fire as a threat to everything they cherished. It had destroyed more than mere vistas. Fire killed people, demolished property, and upended social organization. It was to be fought at all costs.

In 1965, fire was well on the way to becoming a tool, a way to manage landscapes for ecological, environmental, social, and visitor purposes. The failure at Yellowstone and the frustration it entailed suggested a complete regime change in the offing, a new way of thinking about fire. In 1965, this revolution was imminent, in its infancy, but already carrying the weight of science in a culture that saw science as the closest thing to truth.

The change in values between these two instances marked the boundaries of fire management in the national parks. The National Park Service began by trying to control an uncontrollable natural phenomenon, an absurd idea given the size of its domain and the resources available to it. This effort to impose order on nature was doomed to fail; at the same time, the people of the first half of the twentieth century had little choice. They had no viable intellectual mechanism through which to understand nature. By the 1960s, changes in science, technology, and ideology created a different set of ideals. Control of nature was no longer the sole and singular objective of policy. The manipulation of nature through policy had become an accepted goal.

Fire had become a tool, but one that humans had never entirely controlled. This tension, between fire as nature and fire as an instrument, underpinned the National Park Service's approach. Fire had unparalleled power to reshape landscapes, but its use necessitated a level of risk that could spiral beyond the capability of any human entity, much less the NPS, to manage it. National Park Service thinking and action about land management was intertwined with this dichotomy. Nothing was new about this; humans have faced the same dilemma since they discovered fire.

Fire has fascinated, awed, frightened, and frustrated humans since the first human threw a piece of meat into a fire and decided it tasted good. The human ability to harness fire, to set it and to put it out, is unique to the species. Every civilization recognized fire as elemental, basic to human life. They also knew its dangers, the ways in which its power exceeded that of mortals. Fire overpowered humanity at will. It easily escaped the bounds humans established and even under control, it always retained the fearsome potential to destroy all in its path. Fire could burn shelter and belongings, kill people and animals, and destroy the lands they needed to survive.

Though fire is a natural, necessary part of many landscapes, from a human perspective, its power is so great that, unchecked and sometimes uncheckable, it can ruin scenic beauty and productive value. Fire is both threat and asset. Under control, it has been always a valuable tool for humans. Out of control, it can destroy communities, food sources, and ultimately lives. For as long as humans have experienced fire, they have embraced and fought it with equal ardor.

Nowhere has the ambivalent human relationship to fire been more apparent than in America's national parks. These magnificent landscapes and cultural treasures were preserved by congressional legislation and presidential fiat to provide inspiration and succor for a country in search of a distinctive identity. These parks were American, both in their character and their meaning, and they spoke volumes to a changing nation in the late nineteenth century and throughout the twentieth century. Unlike the baronial lands of Europe, the American national parks embodied democratic ideals: they were set aside to be held in common, enjoyed by all Americans in perpetuity. For many, the parks reflected not only the might and majesty of a great nation, but also its highest aspirations: equality, initiative, wisdom, and concern for the future. Fire, in these quintessentially American spaces, proved problematic.

Entrusted with the nation's symbolic heritage, the National Park Service managed lands that were far more important than their size. In effect, fire in national parks served as a barometer of the management of public lands. Fire moved to the center of national debate about wilderness and recreation. When parks such as Glacier and Yellowstone burned, the American public not only took notice, but demanded a response. In that paradox, science challenged historic precedent as the National Park Service led federal agencies toward what has become the standard mode of response to wildfire on federal and private land throughout the nation.

Fire was at the heart of the constant tension contained in the National Park Service's twin goals: to preserve nature—which included fire—and to make nature available for the enjoyment of all Americans, which meant no less than fighting fire and often required excluding it. For an agency that depended on the public's affection to carry it through rough patches with Congress and local constituencies, this was no small tension. The result has been an experiment, a laboratory in which national parks have both brilliantly led and slavishly followed trends in fire management.

As both leader and follower, the National Park Service has been forced to constantly recalibrate its management policies and techniques. The Park Service began, as did all federal agencies, by suppressing fire whenever it could. This strategy bred long-term ecological problems, and yet introducing fire, which the NPS developed to better manage landscapes, required the fundamental reeducation of an anxious public and press. The use of fire as a management tool held the eternal prospect of a blaze that would spiral out of control, destroying property and threatening human life, obliterating decades of planning and preparation with one gust of wind on a particularly dry afternoon.

Since the establishment of Yellowstone National Park—the world's first national park—in 1872, the desire to suppress, control, and eventually use fire has been integral to the management of federal park areas. Following the establishment of Yellowstone National Park, the dominant effort was to suppress wildfires. The idea of complete fire suppression began in the national parks with the appearance of the U.S. Army in 1886, and the model was carried to other federal land management agencies over time. Fighting fire and putting it out when possible was simply the human response to the phenomenon. In most cases, suppression was easier to imagine than to achieve. Under army administration, sincere efforts to put out fires consumed considerable military energy and resources. After the founding of the National Park Service in 1916, suppression in the parks depended on the willingness of Congress to provide money to combat the blazes. The pittance that arrived pushed the infant park service to emulate the U.S. Forest Service, which soon established itself as the lead agency for fire prevention. Forged in the flames of the brutal summer of 1910, the Forest Service treated fire as its mortal enemy. It controlled the vast majority of funding for federal fire response, and its approach dominated for the better part of the twentieth century.

This situation began to change during the 1960s. A series of shifts in science, ecology, and management philosophy contributed to a revolution in the NPS's approach to fire, which became service policy in 1968. The National Park Service became the first federal land management agency to recognize the myriad ways that fire could help to maintain the landscapes so dear to the American public, mirroring the practices of Native Americans before the arrival of Europeans. The national parks became the testing ground for intentionally ignited fire, as well as for experiments in letting natural fires burn. Ecologically sound, this strategy was revolutionary, threatening, and even dangerous, yet the NPS persisted in the face of challenges to its authority and in some cases, intense questioning of its judgment.

Fire became central to every debate about the use of public land in the United States. Its management defined the wilderness and recreation debates. The impacts of extractive resource use and tourism were set against the backdrop of the threat of fire and the prospects for its management, and fire defined the changing relationship between science and ecology. National park lands mattered in this debate not because of their vast size, but because of their emotional importance to the American public. It meant a great deal more when national park lands burned than it did when ordinary forest lands erupted in flames. National park lands embodied national identity; even those far from them who had never seen them felt a twinge when they saw pictures of iconic national parks in flames.

As a result, national park managers faced a more difficult dilemma than did other public land managers. Their lands and consequently their decisions about how to address fire occurred in high relief, visible to every element in American society. Park officials managed in a public environment in a way that other managers did not, for the public was always watching. On this stage, the drama of fire played out in ways that it could not in other venues. The decisions about what to do with fire on national park lands occurred in public, subject to the whim of any newspaper columnist or television commentator who might weigh in on the subject. This complicated the application of scientific principles, for science and public policy were often at loggerheads, and forced the NPS to think about its management objectives as the intersection between science and public relations. In the resulting morass, the national parks became even more important, for they highlighted the intersection of science and popular debate. This did not make resolution or the development of management philosophy and policy any easier.

It took twenty years for the philosophical commitment to fire use to evolve into a formal planning structure that encouraged its introduction. Fire planning covered everything from the response to natural and accidental fires to the rules by which fire could be introduced into national park landscapes and the conditions under which this process could take place. The innovations came slowly, codified in 1978 in NPS-18, the first set of formal rules in national park history to govern the use and suppression of fire. During the early and mid-1980s, its principles were applied in fire plans throughout the national park system. As the decade drew to a close, the NPS had a structure and process for managing fire, albeit one that had yet to be seriously tested.

In the summer of 1988, that test came: the National Park Service faced a major fire at Yellowstone. Though earlier experiments in fire use had gone awry, the consequences had been local. Major fires at the nation's most iconic national park drew a wide set of critics, turning fire management into a national political question. The inability to control the fires turned into a referendum on federal management of land in general. The fires remained in the forefront for most of the summer prior to the presidential election. The result was a challenge to NPS fire policy and objectives that threatened not only the way the National Park Service addressed fire, but also the values at the very center of NPS management.

In response, the NPS reshaped its new fire policy, often guided by the Department of the Interior and pressure from Congress, states, and local governments and culminating in a national fire management plan in 1995. As the 1990s ended, the NPS had redefined its policies and instituted greater safeguards. It faced a century-old problem: much of the land under its care and even more of the acreage surrounding national parks had been subjected to suppression for a very long time. This meant that the lands in question had experienced a tremendous build-up of combustible material—downed trees and limbs, fallen leaves, young trees, and similar plant life. This built-up heavy fuel load characterized national park lands; little of those forests had been treated to limit any of the consequences of suppression. In a climate in which both urban and rural wildfires had become a regular feature, the NPS wisely anticipated destructive fires on its lands.

That expectation was realized in 2000, when a fire on the North Rim of the Grand Canyon and the Cerro Grande fire at Bandelier National Monument provided severe examples of prescribed fires—fires set intentionally for management purposes—that escaped control and caused considerable damage. In both cases, evacuations of communities followed. Near Bandelier is the Los Alamos National Laboratory, and the proximity to the headquarters of the nation's nuclear and weapons research programs exacerbated the danger and fear that stem from any major fire. These fires seemed like errors in judgment, and they led to questions about the efficacy of introduced fire, as well as to concerns about the National Park Service's management strategy.

As the twenty-first century dawned, the old rules of fire management had been shattered by experience. Suppression no longer reigned supreme; the intentional use of fire had been developed, challenged, and then improved by the experiences of a generation of application. Fire had a firm role in the national parks, but the evolution of management in response to changing patterns of living, politics, and statutes remained uncertain.

The public debates about fire in parks continued. A generation after education efforts about the use of fire began, the public still regarded suppression as the ultimate practice. As Americans moved away from the environmentalism of the 1970s and 1980s, their tolerance for fire on public lands and in national parks diminished. Especially when fire threatened or destroyed property, the calls for suppression rose in volume. This strident response became an important part of the debates over the use of fire as a tool in national parks.

The inexorable problem remained: even with thirty-five years of intentional fire, less than 2 percent of the 84 million acres of American national park land had been treated by fire. This meant that suppression had been the operative practice on most land and when any of that land caught fire, the consequences of an extended suppression regime came home to roost. Ninety-eight percent of national park land more than 80 million acres, would burn hot, for its fuel loads had built up over most of the century. Only luck and deft skill kept every summer from being as hot as 1988, 1994, or 2000.

As long as there are national parks, fire will remain a management issue. It is one constant in varied landscapes. The history of wildfire management in national parks has paralleled the evolution of national park management, which in turn has followed the cultural trends of the twentieth century. The increase in categories and types of fire that accompanied the shift to a policy of fire management rather than suppression reflected both the increasing professionalization of the National Park Service and political pressures. Ecological research began to show the ways in which fire could be restorative. After 1968, NPS policy embraced the philosophy that natural fire had to be nurtured where it continued to thrive and fire reinstated where it had been suppressed—except near human habitation or essential infrastructure, where suppression would continue.

This became a matter of practical ecology. It also served as a highly symbolic expression of a change of mission, that national parks should be managed not as primarily recreational or scenic entities but as coherent natural ecosystems. Americans' relationship to the wild had to change from control to celebration of its natural processes. Preserving free-burning fire acquired the same cachet among scientists as preserving grizzlies or restoring wolves; it was a much harder sell with the public at large. New fire terms reflected freshly minted fire policies that in turn articulated new values. This seemingly arcane debate expressed a deeper turmoil over how American society should exist on the continent. Fire had an internal logic, American culture had another; and the two often collided spectacularly in precisely those places, such as Yellowstone, Yosemite, and the Everglades, that had become cultural icons under the aegis of the National Park Service.

ONE

Fighting Fire on Horseback

The Military in the National Parks, 1872–1916

On August 20, 1886, Captain Moses Harris and his fifty-man cavalry troop arrived at Yellowstone National Park, made camp at Mammoth Hot Springs, and took command. Harris was the first U.S. Army officer to find himself the administrator of a national park, an entirely civilian task that was well outside his training. What he found was a park aflame. All over the park, fires were burning, the most severe of which originated just days before Harris's appearance, about seven miles from the Mammoth Hot Springs Hotel.

A pattern began at that very instant. The soldiers and the resources available to them were simply not sufficient to extinguish a fire of this size, and it spread. The men got out their shovels and dug fire lines, hauled water from rivers and lakes, and even initiated backfires, but to no avail. In October, the fire still burned, having grown to as much as sixty square miles. A lesser fire had started along Tower Creek in early September, and a few others appeared and either burned out or were extinguished by Harris's troops.[1] Although soldiers spent a good portion of their time fighting fires during their first months at Yellowstone, the application of military resources was no guarantee of effective fire management.

The creation of national parks occurred in the context of the great changes that swept the nation after the Civil War. With the reservation of 2 million acres for Yellowstone National Park in 1872 amid rampant industrialism and growing economic inequities, the United States declared itself "nature's nation." Nathaniel Pitt "National Park" Langford, the transplanted Montanan who became the leading American proponent of national parks, was on the Northern Pacific Railroad payroll when he visited Yellowstone as part of the Washburn-Doane expedition in 1870. He later lectured with stereopticon images of Tower Fall, the Yellowstone River, and the geyser Old

Faithful. National parks affirmed the ideals of democracy, he proclaimed. While in Europe, kings and barons owned such lands, in the United States, spectacular nature truly belonged to the people. Yet in all the huzzahing and hurrahing that surrounded the national park proclamation, no one gave much thought to its management. Although Yellowstone National Park was assigned to the Department of the Interior, no federal agency received specific authority to manage this vast area; no organization or entity jumped to the rescue to protect the park, manage its many resources, and prepare it for visitors.[2] This dilemma continued to haunt the national parks for the next four decades.

The intrepid Langford was appointed to the unpaid position of superintendent but as U.S. bank examiner for the territories and Pacific Coast states, Langford was occupied elsewhere; he made only three short visits to the park during his superintendency. Langford wrote rules and lobbied for the protection of park resources, but was denied even his limited appropriation requests.[3] He could not defend the park against hunters, intruders, or natural elements. His successor, political appointee Philetus W. Norris, who arrived in 1877, fared little better. In 1878, Congress finally provided a $10,000 appropriation to "protect, preserve and improve" the park, but the futility of that act was driven home that same year, when a group of Nez Perce attacked tourists in the park, killing one.

Once assistant superintendents were appointed, they functioned in the capacities later associated with rangers. Careful with their own fires, they insisted that tourists who camped in Yellowstone show equal vigilance. Almost every visitor to the park built a fire every day, creating the potential for an almost infinite number of fires to burn out of control. In 1879, even though July, August, and September had remained precariously dry after a stormy June, fires remained at a minimum. Superintendent Norris attributed this success to the persistent watchful nature of his charges and to their ability to impress on park visitors the importance of close monitoring of fires.[4]

A crucial principle had been established: fire suppression succeeded most completely when an educational program accompanied it, when park personnel patrolled heavily used areas with regularity, and as long as manpower and water to devote to suppression existed. Luck in the form of regular rainfall and early winters helped, but even at the most rudimentary level, insisting on prevention went a long way toward assuring protection even before federal troops arrived at the park. Despite the damage done by the "wonton [*sic*] carelessness of and neglect of visitors," a sentiment expressed in the 1882 annual report by Superintendent P. J. Conger, by the early 1880s, park superintendents could legitimately claim effective fire management.

Still, by 1880, it was clear that a more comprehensive system of protection and management was necessary.[5]

The real change in park fortunes came in 1886. That year, Secretary of the Interior Lucius Q. C. Lamar, a Mississippian and former Confederate who had assiduously worked for national reunion, became aware of the provisions of the act of March 3, 1883, which authorized the War Department to provide troops for national park protection upon the request of the secretary of the interior. When Lamar asked the secretary of war for help, he began a nearly thirty-year relationship in which the military provided the primary protection for the growing number of national parks and related areas in the United States.[6]

This action brought the cavalry and Captain Harris to Yellowstone in 1886. Harris faced a dilemma when he assessed his management prospects. He understood the origins of manmade fire in the park. Most of the blazes originated near the road between Gardiner and Cooke City, Montana, a heavily traveled track along which many visitors stopped to camp. Harris believed that those who lived near the park, what the officer called "a class of old frontiersmen, hunters and trappers and squaw-men," were responsible for the remainder of park fires. He surmised that they used fire in two ways: to drive game to locations where they could legally shoot animals and to demonstrate their disdain for any kind of government regulation.[7] The lands that became the park had been open land used communally by local people without restriction.

Many "old frontiersmen" were experienced with the use of fire. They learned their techniques by watching Native Americans and found themselves comfortable with an older, more integrated approach to nature. Fire was an important part of the regime, and while carelessness sometimes led to wildfires, from the point of view of an old hand in the woods, it was often worth the risk.[8] Such practices conflicted directly with the basis of military management, which at its core, boiled down to control of natural forces in the same way that in wartime it sought to master opposing armies. The difference in perspectives set the stage for a generation-long controversy in and beyond the national parks that played out between local residents and first, the army and later, federal agencies, such as the U.S. Forest Service (USFS) and the National Park Service.

Despite the actions of the old frontiersmen, visitors constituted the single most frequent source of fire at Yellowstone National Park. Harris regarded carelessness as the cause of their fires. Campers settled anywhere they chose for an evening or longer, their choices governed by the availability of water, timber for fires and shelter, and even, in some cases, game. They acted as if they were traversing the wilderness, not visiting nascent sacred space. Tourists randomly cut timber and left the detritus of their campsites, even the carcasses of recently shot animals. They did not adequately extinguish their campfires, nor were they consistently careful about the ways in which they used fire for cooking or staying warm. The prevalence of fire in so many places close to the main arteries of travel provided further evidence of the origins of fire in Yellowstone. Where there were people, Harris observed, there was fire. The cause and effect seemed clear, and the army responded in a fashion characteristic of its management practices. By 1889, the military segregated travelers into designated campgrounds to better manage visitors and the fires they sometimes caused.[9]

The military officers who served as the superintendents of the park recognized fire as their greatest challenge. Their troops could manage vandalism, serve the growing numbers of tourists, and track and arrest poachers and other violators. Fire posed a much larger threat. Although forestry schools in Europe had begun to think about timber management, fire was beyond their purview. Not only did no such thing as fire training exist in the United States, any large blaze could easily overwhelm the limited resources at the park's disposal. Although no major fires marked 1887 or 1888, officers recognized that their situation was precarious.[10]

In the summer of 1889, the northern Rockies exploded in flames. Yellowstone's new park superintendent, Captain F. A. Boutelle, was a veteran of more than twenty years in the army. He played an important role in the capture of Captain Jack in the lava beds during the Modoc War of 1872, and he brought the forcefulness that marked his military tenure to firefighting. Boutelle emerged as the most prescient of the early park commanders, implementing a comprehensive program to fight fire. Boutelle's men built a system of roads, installed telegraph and telephone wires in the park, purchased new tools to fight fires, and compelled travelers to stay in the campgrounds. When fires broke out that summer, Boutelle showed decisive leadership and garnered attention from important magazines and from national newspapers.[11]

Boutelle found himself a darling of the early conservation movement, with George Bird Grinnell, a leading conservationist and national figure, supporting the superintendent's position. Grinnell helped to found the Audubon Society, the Boone and Crockett Club, and other late nineteenth-century conservation and culture organizations, and he published *Forest and Stream,* a newsletter that became a leading conservation magazine. Grinnell endorsed Boutelle's strategies; the captain "displayed an amount of energy and decision which promises great things for the future of the Park," Grinnell wrote at the height of the 1889 fires.[12]

Grinnell's enthusiasm for Boutelle's efforts reflected more than a decade of elite concern about fire in the West. Harvard professor Charles S. Sargent had included a map of Yellowstone and extensive commentary about fires in the 1880 census in his *Report on the Forest of North America (Exclusive of Mexico)*, published in 1884. The American Forestry Congress of 1882 had targeted fire as a threat to the nation's forests. Several immense, lethal fires had swept the Great Lake states, the most recent in 1881 in Michigan. By 1886, when soldiers arrived at Yellowstone, many sought institutional means to control the outbreaks. The Adirondacks Preserve, established in 1885 in New York, adopted "fire rangering," a technique of using rangers to detect and suppress fire that was developed in Ontario and Quebec around 1885.

A comparative colonial perspective, particularly among the British and French, also existed. The British had created a system of forest reserves in the 1870s, and they utilized continental experience to address management issues. The opening question asked at the first conference among its on-the-ground foresters was whether fire control was feasible and desirable. The question was new; the existing research into timbered land did not address the efficacy of fire as a tool. There were serious disagreements at the conference, but on British and French lands as in the United States, the crucial experiments were conducted by military units.[13]

Grinnell embraced the idea of suppression as a military obligation, emboldening Boutelle. The army's job was to put out fires in the park, and Grinnell insisted that the Departments of the Interior and War devote more resources to Yellowstone. Boutelle had vociferously protested the lack of resources for firefighting, and his complaining incurred the wrath of Secretary of the Interior John W. Noble. His ongoing disagreements with the secretary, while productive in establishing a formal suppression policy and patterns of resource deployment, led to his removal late in 1890.[14]

The military served more effectively as a deterrent than as a firefighting force. Soldiers prevented people from starting fires by restricting their location and by monitoring their activity within the park. In 1892, Captain George S. Anderson, who succeeded Boutelle, reported that he and his men faced countless fires during the season, but managed to extinguish them by a "ceaseless and numerous system of patrols."[15] Anderson's observation led to further emphasis on centralizing the locations in which visitors camped.

Through most of the 1880s, Yellowstone was the sole American national park. Only Mackinac Island followed until 1890, when Congress established Sequoia, General Grant, and Yosemite national parks.[16] Federal administrative control of lands near national parks was extended in 1891, when Congress created the first forest reserves around Yellowstone, which at first received no more direct or immediate resources than had Yellowstone at its establishment. While the forests stood without protection, army troops were sent to the new generation of national parks under the same terms and conditions that propelled them to Yellowstone and faced many of the same issues there.[17]

All three new national parks shared the "big trees," the sequoias and redwoods that propelled preservation efforts in California and proved more difficult to manage than monumental scenery or charismatic animals. Unlike Half Dome or Tower Falls, the big trees were dependent on fire for their survival. Fire burned cavities into them; fires swept around their trunks almost annually when Anglo-European observers first wrote of the region in the 1850s. Suppression as practiced had the combined effect of changing the ecology of the area around the big trees and altering a historic landscape by excluding a primary catalyst of earlier change. No less than leading conservationist and the nation's first chief forester, Gifford Pinchot, noticed the contradiction. When told that area residents had "saved" the Kaweah big trees from fire twenty-nine times, Pinchot wondered aloud who saved them during the previous 4,000 years.[18]

Among the three new parks, Yosemite enjoyed an iconic status by the 1890s that increased the demands on its new military overseers. By 1890, the iconoclastic Scotsman John Muir had become "John of the Mountains," the most famous figure in early nature preservation. California had become a powerful economic engine, with San Francisco as its central city. From a vigilante frontier town, the city developed into the premier economic center of the West. At the same time, California had stepped to the fore in the complicated embrace of Romanticism, measurable science, and antimodernism that so strongly foreshadowed the rise of legislated conservation. For an urban society grappling with a sense of loss that stemmed from rapid growth and rampant socioeconomic inequity, the beauty and serenity of Yosemite epitomized the cost of this transition.[19]

Not just a symbol of wilderness, Yosemite was a real place, beset by serious management problems that pre-dated its national park status. As a state park from 1864 to 1890, it had become the best example of the struggle between preservation and use that so completely dominated early national park history. Americans across the country focused on the region as the locus for their as yet undefined national identity, an emblem of what made the relationship between the American nation and the land it inhabited special. Not only did the establishment of a state park fail to guarantee protection, the cultural meaning of the new park in a rapidly industrializing society brought streams of visitors. The pressure on the new park's resources grew.[20]

As was the case at Yellowstone, human use of fire to reshape Yosemite preceded the founding of the United States. The Ahwahneechee people, who long had lived in the Yosemite Valley, had used fire to transform their world for their own benefit, a practice common among native peoples across the continent. They prized the black oak, a species that thrived on sunlight, for its black acorns, and systematically fired the region to burn pine, incense cedar, and other less hardy saplings. The black oak trees multiplied when the understory of saplings was removed, creating a vision of an open valley. Despite the removal of Native Americans from the park in the early 1850s, the long-term use of fire resulted in relative stasis in the Yosemite Valley. For at least the twenty-year period between 1850 and 1870, the valley floor looked much the same. After Native Americans were removed from the valley, they returned seasonally to engage in their traditional practices, firing the saplings to allow mature trees to flourish and maintaining the rough equivalent of the biology they had created before.[21]

By the 1870s, visitors to the valley floor reported a severe decline in the overall number of trees. Not only had the undergrowth that been the focus of regular burning disappeared, so had the thick stands of timber that had helped to sustain Native American life. Increased plowing and grazing also led to a more open valley. Later scientists attributed this change to the compacting of the soil that accompanied increased agriculture. The vista was remarkably different: instead of the thick stands of black oak of the 1850s, the valley in the 1870s showed open fields and young pines and cedars.[22] The spectacular valley had been altered by the combination of fire suppression and more extensive agriculture and animal husbandry. The same transformation had taken place at lower elevations as well.

As a state park, Yosemite had been consigned to the care of the Yosemite Park Commission, which administered it for twenty-four years before the national park was created. Although famed landscape architect Frederick Law Olmsted was the initial chair, he soon departed, leaving the park in the hands of commissioners who functioned as a development agency, promoting roads and local business interests. The Department of the Interior opened lands along the park's boundary to settlement, adding homesteaders to the odd array of people under the commissioners' jurisdiction. Throughout most of their era, the commissioners worked closely with mining and timber interests, and as a result, considerable acreage moved into private hands.[23]

The commissioners managed from a distance, deaf to the growing number of competing interests on the park's borders. Although they saw themselves as managers of a park, their desire to preserve scenery was closely tied to its ability to make money. Their annual reports reflected their economic interest in the park's ability to generate revenue, noting the growth of young merchantable timber in the lowlands in 1885–1886. The timber provided evidence of a successful suppression regime; without suppression, no young timber would have survived. For the better part of twenty years, this modicum of supervision sufficed even as enthusiasm for Yosemite grew in ways the commissioners did not foresee.[24]

In 1889, the viability of this form of management ended. That summer, a fire swept the famed Mariposa Grove. Its suppression policy left the Yosemite Commission poorly prepared to address fire and led to disastrous consequences. "That most despicable of crimes, forest arson, the result of carelessness on the part of campers or design on the part of sheepherders, turned the surrounding forest, outside the jurisdiction of this commission, into a flood of fire," the annual report of the commissioners averred. "The conflagration at times almost surrounded the Great Sequoia Grove and invaded it at many points."[25]

In one signal event, the inadequacy of the existing system was exposed. The cause and effect became inverted and the presumptive solution, complete suppression, transformed the grove over the subsequent seventy years. The Mariposa Grove fire was the catalyst for the demise of the Yosemite Park Commission and for the arrival of federal troops to administer the park. The fire was widely regarded as final proof of the commission's inept management, and in a changing nation, Yosemite was seen as a sufficiently significant symbol to merit national protection. The commission's existence was already under assault before the fire. The powerful conservation group that surrounded John Muir—including Robert Underwood Johnson, editor of *Century* magazine; Stanford University president David Starr Jordan; attorney Warren Olney, later the reform-oriented mayor of Oakland, California; scientist Joseph LeConte, a University of California professor who shaped science throughout California; Charles Robinson, an artist in Yosemite; and others—attacked the commission and sought to include the Mariposa Grove in the larger national park they planned.[26]

The Southern Pacific Railroad noted the advantages that other railroads gained from conveying tourists to Yellowstone and hoping to bring more visitors to Yosemite, lobbied for national park status at Muir's and his friends' behest. A timely introduction of the park bill at the end of a congressional session led to easy passage, and on October 1, 1890, President Benjamin Harrison legally made Yosemite National Park the responsibility of the federal government.[27]

The leading environmental figure of his time, Muir had strong feelings about fire. Imbued with a sense of the forests as sacred, he adamantly opposed burning, denouncing it as a much more severe waste than even logging. Herders had long fired pasture in the fall to create better growth in the spring. Muir detested sheep and their herders and was suspicious of their motivations. As a result, he opposed the herders' fires, no matter what their purpose.[28] Muir's pronouncements reiterated the characteristic link between bad fires and bad people, a hallmark of the military's suppositions about fire and its management. In his famous account of two fires, he wrote of one roaring through chaparral slopes and upon reaching the top, slipping quietly through the open forest's understory. In Muir's day, the emphasis was on the raging blaze. A later generation read that landscape differently, placing its emphasis on the quiescent burn.

In 1890, Secretary of the Interior John W. Noble created the first general regulations for national park use. He added specific rules for the three new California parks, General Grant, Sequoia, and Yosemite, for they were more heavily used than Yellowstone. Most important was the regulation that made it illegal to "start or kindle or allow to be started or kindled any fire in grass, leaves, underbrush, debris or dead timber down or standing." Anyone who started a fire would be liable for the financial damage it caused, a stiff penalty for the largely impecunious homesteaders inside and outside the parks' boundaries.[29] The secretary's rules enshrined suppression and insisted on individual control.

Timber cutting and fires that resulted from the needs of tourists also created management issues in the California parks. On an 1890 inspection trip, Department of the Interior special land inspector Thomas Newsham discovered that significant numbers of trees had been cut away to provide visitors with better views of Bridal Veil and Yosemite Falls. "Below this, some distance, there are evidences of a recent fire caused by some tourist campers," he wrote Secretary Noble, "but I am glad to say that it did not extend very far." Newsham observed thousands of young pine, cedar, and other trees, which he expected, "if left undisturbed, will soon make beautiful groves over most of the floor of the Valley." Management had become a struggle between present uses and future opportunities: a resurgent forest would overwhelm the valley and pose a fire hazard, particularly on slopes away from the valley proper.[30]

An ongoing debate that centered on fire had already begun, limited by the combined constraints of loose hierarchical structure and absent resources. In 1890, a representative of the U.S. Coast and Geodetic Survey advocated the employment at the Mariposa Grove of a "young, active, sensible, and conscientious Guardian, appreciating what is needed and proud of the responsibility of such a trust, with one or more assistants of similar character, [who] would soon give a sense of security against fire." This conception, ahead of the arrival of troops, became the baseline for management, an objective toward which managers could strive even if they could not attain it. It did not reflect the ongoing reality that sheepherders and those outside the park boundaries neither appreciated nor respected Yosemite, and "acts of spoliation and trespass," as official documents referred to such incidents, continued unabated after the transfer to federal administration.[31]

The arrival of troops at Yosemite in the summer of 1891 transformed the day-to-day operations of the park. When forty-six-year-old Captain Abram E. "Jug" Wood led his troops into the Yosemite Valley, they found circumstances that differed from Yellowstone in one crucial respect. As was typical in many places in the West, local interests perceived economic value in Yosemite and had secured an effective local monopoly. A small cabal called the "Yosemite Ring" controlled the region, and its leaders resented not only federal intrusion, but the efforts of the budding conservation movement, largely based in San Francisco and its surroundings. The efforts of Muir, Underwood, and others made the fate of Yosemite into a national topic of discussion. Wood proved more than equal to the task, carrying out his mandate with the upright aplomb that resulted from his thirty years in the army.[32]

Fire loomed large among the issues that Wood and his men faced. The variety of human uses, increasing visitation, and tension between the Yosemite Ring and the military led to different kinds of fire. Suppression made natural fires more threatening due to increased loads of flammable underbrush; accidental fires caused by tourists posed an even greater threat because of their proximity to inhabited areas; and malicious fires set by opponents of the park and federal administration heightened the danger. As at Yellowstone, most fire resulted from the carelessness of visitors. Yosemite's fire policy became proactive prevention and reactive suppression.

Wood effectively balanced the narrow space between ideals of aesthetic beauty and the economic goals of tourism. The Mariposa Grove had acquired an almost sacred position in the pantheon of the American spectacular, and for the army, protecting it took on added importance. The fires of 1889 set off a chain reaction that prompted the military to take proactive action. In its annual report after the fire, the Yosemite Commission advocated protecting the Mariposa tract by "surrounding it with a border over which a conflagration can not so readily pass." Implementation of this barrier fell to Wood and the military. His men constructed a perimeter road and cleared dead and downed wood in 1892 and 1893. The debris was piled and burned in a systematic fashion, creating an open zone around the grove that enhanced its unique qualities as well as the advantages of the army presence.[33] The military's aggressive action to prevent a repeat conflagration at Mariposa Grove and the growing national importance of Yosemite National Park drew attention to fire as a serious adversary of national parks. Military suppression also climbed a notch, as did the concept of proactive management.

Conversely, the clearing around the grove added another dimension that fit with the values of early conservation. By removing downed trees and underbrush around the big trees, the military contributed to the designation of the area as sacred space, apart from the profane space of human living and industry. This articulation meshed perfectly with the Sierra Club's standard, the idea of a challenging nature without visual human intervention that had great aesthetic value. The pattern that dominated the first century of American conservation had been set, and fire and the army's response to it played a role in reinforcing those designations.

Despite such efforts, destruction of timber in and near the park continued, much of it left lying around, creating a possible fire hazard. Agricultural development within the park brought barbed wire, which further divided the land and limited the impact of military patrols. Some plants diminished in number, further evidence of human impact and the danger of ever-growing settlement in the region. By 1892, the Yosemite Valley floor looked as if it were a "poorly managed cattle ranch," observed General Land Office (GLO) special agent Captain John S. Stidger. The park neither preserved the natural setting nor protected resources from potential calamity.[34]

Fires continued to vex not only Yosemite National Park, but the entire Sierra Nevada region. The 1890s saw a significant change in management efforts, for federal officials, GLO special agents prominent among them, began to visit and review land use practices throughout the Southwest. Homesteaders and ranchers had raised crops and animals without oversight for at least a decade, and they, like other westerners, resented the appearance of federal officials. They viewed fire as an essential component of their lives, something they simultaneously feared and upon which they relied. At the Plumas Forest Reserve in California in 1904, a forest supervisor noted:

> [T]he people of the region regard forest fires with careless indifference. . . . The white man has come to think that fire is a part of the forest, and a beneficial part at that. All classes share in this view, and all set fires, sheepmen and cattlemen on the open range, miners, lumbermen, ranchmen, sportsmen, and campers. Only when other property is likely to be endangered does the resident of or the visitor to the mountains become careful about fires, and seldom even then.[35]

Though later studies of tree rings suggest that fire in the region actually declined after 1864, a battle of perceptions that reflected predispositions about the uses of fire took shape, setting local people and federal managers in firm opposition.[36]

The same tension was evident at Sequoia and General Grant national parks. The Giant Forest at Sequoia loosely mirrored the Mariposa Grove at Yosemite, and the minuscule General Grant served almost as a noncontiguous section similar to the Minaret area of Yosemite. When Captain J. H. Dorst and Troop K of the Fourth Cavalry arrived at Sequoia in the summer of 1891, they found conditions and conflicts that roughly paralleled those at Yosemite, in particular the struggle to protect the park when the state owned the surrounding resources. Cutting of state timber had become an especially dangerous source of fire, Dorst noted, and he advocated the transfer of much of the surrounding state timberland to the park as a way to limit the threat. In his estimation, the state was too vulnerable to local constituencies to provide adequate protection for the park and its resources.[37]

Sequoia and General Grant soon mirrored Yellowstone and Yosemite in their struggles with fire. The army responded to the fires it saw, mostly lightning fires near inhabited areas or those started by people. Backcountry lightning fires burned out of view. A significant percentage of the cavalry's work hours was devoted to containing fires within the parks. Most blazes were small, requiring a single detachment and a few days to bring them under control. Occasionally, fires spread or separate blazes merged and containment became more difficult, but rarely did they threaten property or life. The consistency of fire suggested some level of intentional burning, which in turn led to a closer look at the activities of homesteaders and herdsmen in the immediate region.[38]

In the late nineteenth-century Department of the Interior, the General Land Office took responsibility for most investigations on public lands. Its special agents were among the most dynamic and experienced members of the federal land bureaucracy. Beginning with the Homestead Act of 1862, they investigated homestead, timber, and countless other kinds of claims throughout the West. A GLO special agent was the natural choice to assess patterns of resource use and their impact on federal lands in the Sierra Nevada.[39]

During the summer of 1894, the GLO dispatched Special Agent W. F. Landers to the Sierra Nevada to investigate the causes and effects of forest fires. Landers had no specific knowledge of fire, but he had been a special agent for more than a decade. He knew his way around the West and was especially good at reflecting the views of locals. Landers understood that he had entered a complicated situation where the goals of his agency and those of locals were unlikely to coincide. At the same time, the power of the federal government over western lands was growing. Despite a number of scandals at the highest levels, the GLO had asserted itself as an arbiter of western land issues. Landers's deployment was a characteristic response to the growing questions of the region. It marked a belated effort to bring the experience of the federal government to the complicated situation in the California mountains.

After a summer of observation, Landers concluded that the actions of sheepherders, still prominent in the California mountains, were the primary cause of forest fires. After pasturing their animals on public lands in the summer, sheepherders drove their flocks to lower elevations and in a time-honored practice, returned to the uplands to set fire to trees and meadows to create pasture for the following season. These were hardly unusual practices. Native Americans had engaged in broadcast burns along routes of travel and for resource extraction for as long as they had been in the mountains, and throughout the West, immigrants from the Basque region of Spain and other sheepherders had undertaken similar practices. Although he did not believe cattle were a major source of the problem, Landers discovered that cattle and sheep men in the area had created a rationale for continuing their practices. They firmly believed that forest fires helped rather than hurt the big trees. Landers was less sure that the fires provided a benefit.[40]

Landers's research suggested that fire management in the California mountains was not simply a problem of perception but of practice. Local practice challenged the military construct, formulating the battle between suppression and fire use as a struggle between good and evil. To successfully implement a suppression policy, the cavalry needed to battle fire before it started—to engage in a program of education and dissuasion as it had at Yellowstone and Yosemite. But Sequoia and General Grant national parks presented a new challenge. At Yellowstone and Yosemite, the cavalry was asked to manage land within the parks' boundaries, a task for which its numbers and skills were admirably suited and where its influence was at its greatest. At Sequoia and General Grant, most of the threats to the parks came from outside their boundaries. And many of the culprits only traversed the region seasonally, making an ongoing campaign a far more difficult task. Although troops could manage the parks and make inroads on other federal lands, it could not easily compel changes in behavior outside the parks.

Yellowstone had already experienced major fires, but Yellowstone's fires were very different from those in California's Sierra Nevada. Typically, Yellowstone experienced crown fires, fires that burned in the tops of trees, through lodgepole pine, except in the Lamar Valley and similar winter ranges, where fires burned through grasses and shrub. The Sierra Nevada experienced surface fires, which spread on the ground, through various pine, chaparral, and fir complexes. Army officers inaccurately applied the lessons of Yellowstone to the Sierra Nevada, suggesting a disadvantage in having a single agency manage two ecologically different parks.

This transfer of knowledge proved inadequate. In August 1898, an extensive forest fire spread throughout the northwest section of Sequoia. A combination of state forestry agents and the cavalry had little success containing the fire, and it spread wildly until it burned itself out in late August. Although the fire did not harm the Giant Forest, the grove of sequoias that gave the park its name, it did introduce a new fear of fire in the region. A state forestry agent was injured fighting the fire, the first known case at Sequoia. The next year, two more fires burned out of control in the northwest part of the park. Both started outside Sequoia and appeared to be intentional, presumably set by herders.[41]

Two summers of major fires forced the army to change its park fire management policies. The failure of suppression meant that existing policies would have to change, and the circumstances provided a way to apply new goals to park lands. In 1898, J. W. Zavely, a GLO special investigator who served as acting park superintendent, used the fires to exclude 20,000 sheep from the park and to remove any remaining sheep from General Grant National Park as well. With herders as the culprits for most fires, this move not only improved the ecology of the parks, it also removed a major source of human-induced fire.

Zavely's bold move was an initial step that eliminated only half the problem. It did remove animals from the parks, but it could not address actions that took place outside of the park but affected its resources. Captain Henry B. Clark, the new acting superintendent, continued Zavely's policy, asking to extend his troop's stay in the park until November 1, 1899, in an effort to combat both fires and trespassing hunters. Together, the two made protecting the large trees the park's primary firefighting priority.[42]

As the nineteenth century drew to a close, the army established a model of firefighting that resembled combat. The army saw fire as an adversary in the same way that it regarded a foreign army. This model would be sorely tested in 1910, aptly titled the "year of the fires." That summer, the inland Northwest erupted in flames, the result of the meeting of lightning sparks, locomotives, scattered humanity, and heaps of burnable fuel left behind by railroad and town building, mines and logging. The previous wet winter, the subsequent dry spring, and a drought-like summer exacerbated conditions. A fire of epic proportions worsened as the summer passed until finally, the Big Blowup of August 20–21, 1910, in Montana and Idaho consumed towns, villages, railroads, mining camps, and anything else in its way.[43]

Two national parks, Yellowstone and Glacier, established in May 1910, stood in the path of these fires. The army had administered Yellowstone for almost thirty years and listed fire suppression as one of its three main obligations at the park; in most years, its forces controlled fire with brigades of men wielding picks and axes. During some prior years, such as 1901, fires burned beyond the capability of the army. That summer, *Forest and Stream* reported that "axes and shovels were the only weapons of use. . . . water buckets are the best 'side arm' a soldier can carry." Despite assistance delivered by troops from Fort Keough, Montana, the blazes burned throughout the region until fall rains brought them to a close.[44]

Fire attracted the attention of the park's most renowned early chronicler, Hiram Chittenden, the engineer in charge of building Yellowstone's road system. His *The Yellowstone National Park* had become the most widely read book on Yellowstone, the source of the considerable amount that the public knew about the park. By the 1905 edition, Chittenden recognized that fire was a primary park issue, a "source of anxious solicitude" for its military administrators. "The control of a forest fire," he wrote, "is next to impossible except by the aid of rain." Chittenden advocated a proactive program of fuel load management, breaking up dense masses of vegetation, the accumulated fuel load created by time and successful suppression, but he opposed using fire as a tool to accomplish this end. Despite Chittenden's recommendations, little was done to lighten fuel loads, and Yellowstone remained vulnerable.[45]

In 1910, Yellowstone's timber went up in flames along with the rest of the inland Northwest. Lightning far from the main roads ignited most of the park's fires, and high winds spread the blaze. A large area south of Yellowstone Lake burned, and new fires erupted throughout August and into September. In early August, more than 200 soldiers battled fire in the park. Despite adverse circumstances, they succeeded in stopping at least two of the outbreaks. Another fire remained out of control until a shift in the direction of the wind drove it toward Yellowstone Lake. Even as they fought fires, soldiers continued their preventive measures. Assiduous patrols of campsites helped to keep new wildfires from starting, although at least four began when fires that were not entirely extinguished transformed into wildfires before the army reached the scene. Fire cut off one troop of twenty-nine soldiers and backed them up against the lake. They managed to fight off the fire and escaped with a few burns and a considerable loss of equipment.[46] By all accounts, the soldiers performed valiantly, but their efforts usually had little direct effect on the fires.

Glacier National Park provided a different kind of challenge. Established in May 1910, it lacked the history of fire management experienced by some of the other national parks. In addition, Glacier National Park was in the middle of being transferred from the Forest Service to the Department of the Interior when fires broke out there in 1910. Just as the fires began, Secretary of the Interior Richard A. Ballinger sent an inspection team to Glacier National Park. Typical of the survey parties sent out to assess land in the late nineteenth century, it contained scientists, officials from nearby national forests, and Ballinger's personal representative, the chief clerk of the Department of the Interior, Clements Ucker. The fourteen men and their ten pack animals entered the park and found themselves in the middle of a maelstrom. For an entire week, the fires prevented their communication with the outside world as a frantic Ballinger tried to reestablish contact. When Ucker extricated himself and his party and reached Fort Yellowstone on August 12, he announced that the park was a "veritable fire-trap." He announced that the Forest Service had done nothing to prepare for the fire season during its long tenure in the region—despite its desire to show the world that it could control fire.[47]

In fact, the Forest Service did no less at Glacier National Park than it did anywhere else in the inland Northwest. Fire simply spread beyond the capability of diverse and poorly staffed agencies. At least 2.6 million acres of national forest land burned in the northern Rockies and an additional 2.4 million elsewhere, and certainly much more land that was not counted was burned as well. The fledgling Forest Service was not equal to the task. At the time, individual foresters administered as much as 1 million acres, often by themselves. They possessed small budgets, with little access to additional resources. When foresters needed help, they recruited workers from local and regional populations. Sometimes, the agency went to cities in the region, such as Spokane, Washington, in order to find people to fight fires. By August, the agency had more than 5,000 firefighters on its payroll, but the number was nowhere near enough to stop the fires. Nor was the available technology, essentially shovels, axes, and water hoses, equal to the conflagration.[48] By any legitimate measure, both the Forest Service and the army performed admirably in their response to the fires of 1910.

Ucker's indictment gained attention in the regional press because of presumptions that the fires, in Glacier in particular, resulted from human malice. The Great Northern Railway had laid off a sizable number of workers earlier in 1910, leading some to contend that a combination of the newly unemployed and wayfarers had started fires in order to secure work putting them out. The accusation contained some truth. Instances of individuals igniting fires and then landing on the firefighting payroll were common, but as an indictment of Forest Service policy at Glacier, the argument lacked credence. The most dangerous and destructive fires in the summer of 1910 started by lightning and grew in force and size precisely because they were far from human view.[49] They were complemented by a welter of human-induced fires, many caused by sparks thrown from railroad cars in the dry climate, and an array of other incendiary events that tied up resources and contributed to the breakdown.

Despite such realities, the fires at Glacier National Park became part of the battle between Ballinger and Pinchot, the head of the Forest Service. At its core, this was a power struggle over control not only of timber but of the very values of conservation. Pinchot advocated suppression at this time, and Ballinger and his agency opposed anything that the former "crown prince of the Roosevelt administration" wanted. Although Pinchot was ousted from the Forest Service, and Ucker's subsequent belligerence was widely acknowledged, Pinchot had achieved control of the terms of the struggle. Not only was he far more adept at public relations, he was acknowledged as a leader in scientific forestry. Even Ucker acknowledged that reality, calling on the Forest Service and the much overtaxed army to respond to the crisis. The expenses associated with fighting the fires figured into his call for the Forest Service. Glacier's entire appropriation during its first year of existence was a mere $15,000, a sum so small that its application to combating the fires would have exhausted it in a matter of days. In contrast, the Forest Service appeared willing to carry the costs of firefighting until Congress agreed to a special appropriation after the end of the fire season to cover all expenditures.[50]

By early August, the fires were so overwhelming that Henry S. Graves, Pinchot's successor at the Forest Service, asked for the assistance of the army at Glacier. Among the troops sent to the park was Company K of the all-African-American Twenty-Fifth Infantry under command of Lieutenant W. S. Mapes. While other soldiers in the park found themselves with difficult but manageable tasks, Company K found itself doing the most difficult work. With two gangs of additional men, a thirty-seven-man crew of lumbermen hired by the park, and a thirty-five-man crew of Greek laborers offered by the Great Northern Railway, the army company had to battle the most powerful blaze in the park. Racial tension and differing goals complicated the interaction. Some of the hired crews refused to work more than a regular 7:00 a.m. to 5:30 p.m. schedule. The untrained laborers were only marginally useful until Lieutenant Mapes sandwiched individual workers between the soldiers. Disciplined troops outperformed the less trained workers in the brutal tasks associated with containing fires.[51]

In the end, the combination of an on-the-ground presence of even rudimentary firefighters and a willingness to absorb costs until reimbursement gave the Forest Service control of the battle against the extraordinary fires of 1910, and ultimately over the culture, policy, structure, and organization of firefighting on public lands. Despite the acknowledgment that Forest Service efforts in the national parks and elsewhere amounted to little that brutal summer, only the Forest Service appeared ready to shoulder the enormous burden of firefighting in the West.

In the two affected national parks, the results of the fires were devastating. At Yellowstone, the fires burned more than 60,000 acres, and firefighting efforts drained the park's limited budget. The tourist hotels were not threatened. The military paid day laborers a total of $12,550 to help the troops dig fire lines and even tried backburning, starting fires to burn into the major conflagration as a way of stopping its progress, at one location, but their actions did little to slow or stop the fires. By 1911, the park had begun to construct fire lanes, but the military had too few people and too little experience to manage the national park, fight fires, select salvage timber for sale or disposal, and maintain order.[52]

At Glacier, about the same area, 60,000 acres, burned, and there the ability of the Department of the Interior to support its national parks was thrown into question. Much of the area had considerable value as a source of timber, but in park supervisor Major William R. Logan's estimation, little scenic value. The major fires were away from the areas frequented by tourists, but many were adjacent to the railway. The expenditures associated with the fire were astronomical, and the Department of the Interior had little but the park's basic appropriation, primarily allocated for road building, to cover its costs. Before the establishment of the National Park Service, individual parks received direct appropriations that were far from generous and usually earmarked for specific purposes. In the case of a cost overrun, as at Glacier, the Department of the Interior needed to know what its expenditures had been in order to request an additional appropriation from Congress.[53]

Glacier National Park provided a flashpoint for the tensions that would come to revolve around fire. Two important federal agencies, the army and the Forest Service, grappled over control of an important dimension of land management with enormous implications for national parks. The army pursued a mode of suppression derived from its experiences in other national parks since 1886, a pattern the Forest Service followed as it carved its own way in the world of land management. The Forest Service tried to rely on military help to fight fires in the national forests, extending the pattern begun in the national parks and firmly locking the suppression mandate of the army in place, but the War Department would not allow the deployment of troops in the national forests. The Secretary of War determined that his department lacked the staff to support such an enormous endeavor.

Elsewhere in the national parks that summer, fire problems were minimal. Mount Rainier and Yosemite both experienced a number of fires, but they paled in comparison to the ones in the inland Northwest. At Mount Rainier, the only fires that required action were the result of unattended campfires. Yosemite experienced a number of fires, including one that burned within a half mile of the Mariposa Grove, but only that fire required the attention of troops. At Sequoia, the only notable fire resulted from blasting on a road project, and Wind Cave and General Grant both experienced typical lightning strikes.[54]

Ultimately, the 1910 fire season proved to be pivotal. Founded in 1905, the Forest Service discovered its purpose that brutal summer. Fire became its religion, the way in which the agency defined itself for the next half century. Nothing was more important to the culture of the Forest Service than fire. The Forest Service derived power from its struggle in the summer of 1910 and defined the terms of the battle against fire. After that summer, national parks followed the Forest Service's lead in managing fire, and for the better part of the subsequent fifty years, suppression dominated that strategy.

Suppression had its vehement opponents, most notably California advocates of "light burning." This practice, the regular burning of surface underbrush and litter, sprang from the conviction that routine burning had produced the forests, kept fuels down, and prevented larger fires. Also called the Indian way of forestry or, in a pejorative variation, "Paiute forestry," light burning had been advocated in California as early as the 1880s. Settlers and timber owners saw light burning as a sure way to reduce fuel load and limit uncontrollable conflagration. As early as 1902, calls to cease total suppression emanated from ranchers and timber companies in the California mountains, leading to a struggle between federal representatives, at this time mostly the army, and settlers.[55]

Light burning gained enthusiastic endorsements in 1909 and 1910. T. B. Walker, a timber owner near Shasta, California, had been a proponent of light burning for more than a decade. Although federal managers in the Departments of the Interior and Agriculture pronounced the practice ineffective for large areas, Walker published an article for the National Conservation Commission in 1909 that described his practices. Another Shasta resident, G. L. Hoxie, a self-described timberman linked to the Southern Pacific Railroad, advocated mandatory light burning the following year in a piece in the influential *Sunset* magazine. Hoxie's call was the most radical yet, but it came just as the worst of the fires of 1910 broke out.[56]

The light burning controversy provided a focus for Forest Service goals. The agency was devastated both by the fires of 1910 and by the dismissal of Pinchot, and it needed a new focus. Light burning represented a collection of practices that were the opposite of Pinchot's vision of systematic, scientific national management of resources. Even worse from the Forest Service's perspective, the hated Ballinger had advocated light burning. The Forest Service revamped itself as a firefighting agency, its commitment to suppression and its contempt for light burning complete. Despite some efforts by Pinchot's successor, Henry Graves, to experiment with light burning, the Forest Service focused its newly prodigious fire management expertise against light burning.[57]

After 1910, a series of changes in conservation culture created a powerful impetus for the creation of an agency to manage the national parks. Military reluctance to continue in the role of national park management, an arrangement terminated by the secretary of war on May 1, 1914, also increased the obvious need for some kind of system for park management. With war looming in Europe and the United States involved in an expensive incursion into Mexico, Secretary of War Lindley Garrison determined that the Department of War would no longer pay for the management of national parks, which he believed should be paid for from appropriations for public lands rather than from the military budget. The nearly $400,000 per annum from the military budget for national parks seemed to him an "abuse," and he served notice that it would no longer continue. National parks had begun to be seen as reflections of the essence of American nationalism. These factors combined to open the way for the passage of the Act to Establish the National Park Service, which President Woodrow Wilson signed on August 25, 1916.[58]

Because of this new law, the final Progressive Era federal land management agency was born. Labeled a "service" with the objective of serving rather than managing as were so many of its peers in that era, the National Park Service was born with a need to establish itself and its position among peer agencies that overlapped with its mission and its constituency. Its primary rival was the Forest Service, and until 1945, the two agencies grappled with a venomous consistency in nearly all endeavors. Such a rivalry reflected both the parallels and the differences between the two agencies. Very often, they offered different plans and programs for the same tracts of land. Their leaders learned to resent each other, and a tenor of distaste often pervaded interagency interactions through World War II.

Given this relationship, it seems surprising that the National Park Service would accept Forest Service leadership in any area, but when it came to fire, the Forest Service led. After 1910, the Forest Service invested significant resources in fire suppression, creating a culture that became the model for federal fire response. It embraced the military ideal of suppression, shaped in the national parks. Once the military withdrew from the parks, there was no other body of federal work power handy. No matter how National Park Service leaders felt about the Forest Service, they had nowhere else to turn for information, technology, and resources to fight fire. The degree of danger posed by fire trumped all other concerns, providing an early model of interagency cooperation. At the establishment of the National Park Service, the army-based system of firefighting was crumbling, and the new agency faced a monumental task. Not only did it have to build an infrastructure for the park system, it also had to fight endemic fires and to resist episodic colossal fires.

TWO The Development of a Fire Management Structure

Fire was the furthest thing from the mind of Stephen T. Mather, the businessman turned civic exemplar who took the reins of the National Park Service at its birth. Born July 4, 1867, Mather was raised in California during an exciting era. He graduated from the University of California at Berkeley in 1887, a devoted adherent of the fraternity Sigma Chi and among the many in the era who inhaled the heady fumes of public service that had begun to entice the privileged classes. Mather first worked as a journalist under the storied Charles A. Dana at the *New York Sun*, leaving the newspaper to seek his fortune in the development of American borax. He took charge of promotion for borax king Francis Marion Smith, creating the Twenty-Mule Team brand and making a name for himself in the nascent public relations and advertising community. Mather had a knack for publicity and for keeping a product in the national eye. Following a nervous breakdown that briefly institutionalized him in 1903, Mather left Smith's empire to join an old friend in a new borax-mining endeavor. Eleven years later, he had become sufficiently wealthy to retire and pursue his other passion: public service. He discovered the national parks and became instrumental in creating an agency to manage them.[1]

Even before he became successful in public relations, Mather had been known as someone with a strong desire to accomplish civic goals. An inveterate joiner, he belonged to nearly every major civic, social, and charitable organization in every city he inhabited. In love with California and among the state's greatest cheerleaders in the early twentieth century, Mather's passion extended to the outdoors, so remarkable in his native state.

Mather's involvement with national parks began with an apocryphal story that illustrated much about the cozy nature of the American ruling class at the start of the twentieth century. One of Mather's college classmates, Sidney Mezes, advised President Woodrow Wilson, leading to the appointment of a coterie of Californians—and indeed University of California graduates—to high posts during Wilson's presidency. One of his contemporaries at Berkeley, Franklin K. Lane, became secretary of the interior. Acquainted with the new secretary, Mather wrote him a scathing letter after a visit to the national parks. Lane purportedly responded: "Dear Steve, if you don't like the way the national parks are being run, come on down to Washington and run them yourself." Whether it happened this way or not, Mather accepted an offer from Lane, coming to the Department of the Interior in 1915 as a special assistant with national parks as the primary part of his portfolio.[2]

Eighteen months later, Wilson appointed Mather as the first director of the new National Park Service. Mather had come to see American nature as the extraordinary characteristic of the nation. Leading the National Park Service offered him a way to illustrate the importance not only of the physical continent, but of the principles of Bull Moose Republicanism, Teddy Roosevelt's political platform in the 1912 campaign, which Mather thoroughly embraced. By 1916, Mather was a dedicated Progressive, among the many who believed that government could and should provide for the common good in the United States.

The idealistic Mather inherited a complicated reality. Little direct management of the national parks had taken place since the army renounced its commitment to the task in 1914. Mather had a very limited staff in the nation's capital and few people in the field. Nor did he have a plethora of resources available to him. Chosen because he had an extraordinary way with people, voluminous connections, access to resources, and unmatched dedication, he faced a task equal to his prodigious skills.

Mather was not a scientist, and fire management was very low on the service's initial list of priorities. He held that flood control, predator removal, and fire management were all difficult responsibilities best left to specialists. Mather had to sell a new idea, that the national government should manage public lands for the enjoyment of the people, to Congress, arrange for funding for park management, build a national constituency for the National Park Service, and otherwise establish his infant agency as a player in the federal land management bureaucracy.

The founding of the National Park Service in August 1916 heralded a new era for both the national parks and national monuments. While Congress had established the initial national parks—Yellowstone, Yosemite, General Grant, and Sequoia—with specific objectives in mind, the passage of an Act for the Preservation of American Antiquities, known as the Antiquities Act of 1906, allowed the president to establish national monuments with the stroke of a pen. The act placed few restrictions on what kinds of sites could be included in the monument category and neglected to provide resources to administer the new areas. The result was a numerical explosion in park areas after 1906, complicated by the remarkable diversity of the new areas preserved. Not only did the park system contain spectacular and expansive natural areas like the first national parks had been, it now included diverse properties that held natural and cultural treasures.[3]

But creating an entity to manage national parks and arming it to fight fires were two different propositions. From 1916, when the agency was established, to the beginning of the New Deal in 1933, the National Park Service lacked the resources to deal with any major conflagration. Park managers rarely had budgets to fight fires; nor did they have the work power or equipment adequate to the challenge of a major fire. As a result, park managers were forced to become beggars for workers and materials, to engage in cooperative arrangements with public and private neighbors, and ultimately to worry every night during fire season if an outbreak of flames might very well be the one that destroyed their entire operation. This circumstance resulted not from any dilatory behavior on the part of agency leaders; instead, they had many demands on their very limited resources, and fire was both so large and so remote, until it actually happened, that it was easy to simply hope it would not.

The fires of 1910 had begun to fade in public memory, and when Steve Mather looked across the federal bureaucracy, he saw the Forest Service actively embracing fire suppression as the agency's core mission. Under Gifford Pinchot's successor, Henry Graves, the Forest Service considered fire protection to be 90 percent of the agency's mission. Faced with his fledgling agency's many needs, Mather likely never considered any kind of proactive fire management regime. His official correspondence shows evidence only of reactions to crises. Despite the incipient rivalry between the Forest Service and the National Park Service, Mather was content to leave the difficult obligation of firefighting to the Forest Service.[4]

Even that agency was only beginning to learn about the enormous challenge it had chosen to make its own. The response to fire remained limited, and the Forest Service recognized that it could do little about major conflagrations. The lack of technology for transporting vehicles and pumping water, the vast array of federal open space in the West, and the incredible need for dollars and labor to make any kind of a dent in a major fire—all simply were beyond the reach of any federal agency throughout the 1920s. To its credit, the Forest Service thought long and hard about fire. Foresters continued to experiment into the 1920s with light burning and with allowing some natural fires to burn, but the Forest Service became primarily a firefighting agency as it built the disasters of 1910 into its creation myth.[5] In contrast, the early National Park Service invested much less of its intellectual energy and fiscal resources in fire. No staff members had fire control or fire management in their job descriptions.

The nascent NPS made early tactical mistakes. Its efforts to attract the public had the adverse effect of increasing the risk of fire. Since the events in Yellowstone in 1886, anyone who managed a national park area recognized that people caused most of the fires that required an organized response. Under Mather, the service's goal was to attract visitors, often as many people as could arrive by the conveyances of the day. NPS efforts were devoted to ensuring that people enjoyed access to exciting natural and cultural sites, were comfortable in the parks, and brought home souvenirs that memorialized their stay. In contrast, the Forest Service was prepared to deny access to its lands as a way of controlling people and with them, the fires that the agency so dreaded. As Mather built a constituency for the national parks, he inadvertently encouraged precisely the conditions that led to fires. Fires outside the backcountry grew in direct proportion to the increase in visitation, but the Washington office of the National Park Service looked past the issue.[6]

Especially in the Sierra Nevada and in the northern Rockies, individual park managers found themselves responding to fire each summer with whatever means were available to them. Problems with communication, transportation, and the inevitable lack of resources placed park superintendents alone in the agency in their battles against fire. They turned to local residents and other federal agencies in their immediate vicinity because these were neighbors with whom they shared problems. Park managers became more than simply national park officials. They took on a wider regional role, developing integrated systems in support of their objectives. In this respect, individual National Park Service units achieved a level of autonomy that they were not generally permitted in more mundane affairs.

As a result, fire remained a local issue for the first decade of the National Park Service's existence. Washington expected individual parks to deal with fires that occurred inside their boundaries or in their immediate vicinity. Superintendents reported fires and how they dealt with them, but they were largely on their own in developing strategies to offset the lack of firefighters and dollars. Superintendents cultivated local residents and worked with area Forest Service rangers to develop response systems. In most cases, the response was positive, since everyone recognized that survival at the edges required interdependence. Only full-fledged cooperation offered the chance of retarding the destructive progress of a blaze.

Even after the end of World War I in 1918, the resources available to the service did not grow in proportion to its obligations. New parks and larger visitor numbers spread limited resources ever more sparsely across the national park landscape. While Stephen Mather and Horace Albright worked to increase appropriations, the pressure on national parks, newly iconic for the well-off public of the 1920s, increased. Despite the popularity of the parks and their managing agency, the NPS budget did not allow for comprehensive management. Mather often successfully secured special appropriations for projects in national parks, but such a strategy served as a piecemeal solution to a many-faceted problem.[7]

As a result, the NPS had to fight fire with funds from its general appropriation. Unlike the Forest Service, which received congressional authorization in May 1908 to overspend its budget on firefighting, National Park Service officials had to divide their meager funds among worthy projects of all kinds, knowing full well that a major fire would require that they shelve projects to pay its cost. Without a special pool of money to fight fires, fire protection and preparation remained an afterthought. The NPS could barely staff its units, and few in the service, save Colonel John R. White, who joined the agency as a ranger in Hawaii and later served as superintendent at Sequoia and General Grant national parks for more than two decades, had any experience with fire or the inclination to think about it.[8]

The dominant technological advance of the age posed another kind of problem. At the founding of the National Park Service, railroads were critical to the service's formulation of its future. Rails would bring the visitors who elected the officials who funded the park system, and their ability to reach national parks topped the early National Park Service's list of goals. Major national parks such as Glacier and Yellowstone were well served by the steel rails either within the parks or near their boundaries. Yet trains and the sparks they threw off remained a prominent cause of fire. In some situations, such as the inland Northwest during a dry summer, a railroad as much as fifty miles away could represent a serious threat to a park. Fires accompanied rail lines with a certainty that was frightening, and park superintendents watched nearby rails with trepidation.[9]

Parks faced other kinds of fires, natural and human-induced. When lightning ignited a powerful fire in a remote area, it required less reaction from the National Park Service. Such fires simply burned until they consumed all the available fuel or were extinguished by precipitation or blocked by geographical barriers. When fires were closer to people, often sparked by human carelessness, parks relied on fear of the fire's spread to catalyze an organized response. Everyone in the fire's path tended to pitch in to slow, divert, contain, or suppress fire.[10]

In its early years, the National Park Service failed to develop an overarching fire policy of any kind, and as a result, the response to fire varied from park to park. Those with considerable timber and significant histories of fire devoted greater resources to thinking about fire, though most lacked the money to make a substantial investment in any kind of prevention program. But most park superintendents and custodians addressed fire as if it were an unexpected event instead of an endemic condition. The most common agency response to a major fire in those years was to look to the skies and hope for rain.[11]

From August 1916 until the summer of 1919, the National Park Service was simply fortunate: no major blazes took place inside national park areas. The small fires in the parks were handled by the zealous application of every available resource. Mather was able to promote the parks, build support for them, deliver influential people to a range of areas, and otherwise further the objectives he and Secretary Lane established without ever facing the reality that hounded the era's Forest Service managers: a fire would expose both a lack of control of the physical world and the elaborate fiction of protection from which federal agencies derived their significance.[12]

In 1919, the first indication that the National Park Service strategy would not work much longer became evident. By then, Glacier National Park had developed a cumulative fire problem. Each year since its establishment in 1910, the park in northwestern Montana faced fires, sometimes large ones. By the end of the decade, northern Montana was full of timber operations small and large, and many had been mechanized. Railroad lines that traversed the region, from large operations like the Burlington Northern to simple log-hauling roads, increased the possibility of incendiary incidents. Sparks from an engine or any of the rail lines could ignite dry brush, posing an ongoing and widening threat.[13]

As activity increased, the number of acres of timber lost to fire rose each year, as did the cost of suppression. In 1917, fire consumed more than 7,000 acres of park timber, and firefighting costs totaled $11,968. Two years later, more than 50,760 acres of timber burned, at a cost of $46,000 from the park budget. Most of the fires resulted from railroad sparks, even though the Great Northern Railway aggressively cooperated to minimize the threat. In 1920, the cost rose to $81,849, a sum that exceeded the park budget for the year.[14]

Superintendents looked at their predicament and the absence of a budget and recognized that creativity was essential if they were to combat the growing problem of fire with any effectiveness. Technological solutions were at the top of the National Park Service's list. By the early twentieth century, some rudimentary fire equipment had become available. In the early 1920s, the famous Pulaski hand tool, the vaunted combination of a mattock and an ax, remained the standard firefighting tool, but a revolution that relied on more complicated pumping equipment already had begun in earnest. The Forest Service tested pumping equipment—developed for urban firefighting—in the field, and Canadian timber companies developed nominally "portable" pumps. Western parks in Canada adopted the technology, and it soon crossed the border to Glacier National Park. The machines pumped water from an available water source—a stream, river, or lake—and horses pulled them from place to place. National Park Service officials eagerly pursued the idea. Superintendent Walter Payne at Glacier enthusiastically endorsed the equipment. Properly applied, he believed, the pump would diminish the impact of campfires that got out of control.[15]

The technology had severe limits; horse-pulled engines required more than simple storage. The utility of the machine depended on a corral full of horses and nearby pasture, close enough to quickly round up the animals when the call came, but large enough to allow them to graze and run to stay healthy. Without the animals in the early automotive era, crews could not bring the equipment to a fire in a timely fashion. Payne asserted that even these rudimentary pumping engines were an improvement over anything the park possessed. He hoped the service would purchase enough to supply a number to Glacier National Park. The NPS could not provide equipment in 1920, but officials planned to seek an appropriation for the following fiscal year.[16]

The National Park Service found that firefighting demanded cooperation with nearby national forests. The Forest Service had long promoted such cooperation, encouraging private timber protective associations in the northern Rockies and supporting other ways to assure a strong response to conflagrations. Other federal representatives, especially the area's grizzled foresters, contributed resources and expertise every time a major fire started. This powerful presence helped to negate the sometimes halfhearted enthusiasm of local residents. Park superintendents recognized the interrelated nature of firefighting and followed the Forest Service's lead, even as the two agencies grappled for control of federal land management. Superintendents complimented their peers at the Flathead, Blackfeet, and Lewis and Clark national forests.[17]

After a significant fire at Glacier in 1921, the National Park Service did as much as it could to develop a fire protection system there. Consistent recommendations emanated from the park, asking for more rangers for patrols in the summer, advocating the construction of fire patrol and boundary trails, and employing fire guards in especially bad years. By 1922, lookout cabins were on the priority list, as was a cross-park telephone line to alert people in case of fire. In 1923, Glacier built telephone lines to Huckleberry Mountain, Riverview Mountain, and Indian Ridge—now called Numa Ridge—and established lookouts. In 1925, the park purchased twelve pumps, its first genuine capital outlay for fire protection, and the new equipment proved to be an effective response to the ongoing fire problem.[18]

Still, each year, at the end of the fire season, national parks had to present their bills for firefighting to the Washington office, which sought special congressional appropriations to restore much-needed operating dollars to the parks.[19] Only in 1922 did the National Park Service receive its first direct appropriation for firefighting, a $25,000 lump sum for the entire system. The service's budget had remained small. Since the founding of Yellowstone, Congress had errantly presumed that the parks would pay for themselves. This assumption made asking for a base budget, much less an additional appropriation, a difficult task. The sum of $25,000 seemed an inconsequential amount after the spate of recent fires, but it paralleled what had become a congressional pattern of after-the-fact spending to solve specific problems in the parks. In 1920 alone, when the entire NPS operating budget was $907,070, Congress appropriated $25,000 for fire prevention to be added to Yellowstone's base budget of $250,000. That same year, Glacier National Park received $62,000 in deficiency spending above its regular allotment of $85,000. Even as the NPS operating budget grew to $3,027,657 in 1925, the $25,000 allotment was a congressional stopgap, a way to appear to address a growing problem and to eliminate the consistent parade of deficiency spending requests. The limits on it were clear: the money could only be used to actually fight fire, not to engage in fire prevention strategies, practices, and tactics. Inadequate even in a year with few fires, the sum represented an important first step.[20]

The $25,000 sum became the standard allocation during the rest of the first half of the 1920s, an annual addition to the NPS's small budget that allowed the service to mitigate at least some of the financial impact of fire. In most years, it did not cover the NPS's total expenditures for firefighting, and large fires still necessitated specific after-the-fact legislation to recoup costs. However, a line-item appropriation in the annual budget represented a different way of thinking about the impact of fire on the park system.[21]

The small national appropriation assured that fire remained a local responsibility, fought with the existing budget and with whatever work power a superintendent could cajole from the surrounding area. NPS rangers often fought fires outside park boundaries, not only to keep fires away from the parks, but also to create a community of interest that allowed park officials access to non-NPS firefighters when a park needed their help. The agency also paid cash for outside firefighters, but finding resources was a difficult process that invariably drained the pool of money for other park activities. Maintenance on roads and trails halted when blazes demanded workers, and the dollars allocated for such activities disappeared into the smoke of any major fire.[22]

The lack of resources devoted to fire complicated an already difficult situation. In the 1920s, the service had already divided fire management into two competing models. The first, fire exclusion, required suppression and was the standard for federal agencies; the second, fire use, involved light burning, but this method had already been relegated to the fringes. In developing a dominant ethic, the service could look to its own heritage with the army, or to the Forest Service, which had absorbed and remade the military example. At the local level, some superintendents departed from the primary model and employed light burning. Officials might selectively let fires burn, as much a result of the lack of funds for firefighting as any ideological reason.

At the national level, the NPS naturally leaned toward the army–Forest Service suppression paradigm. To the early National Park Service, this policy made considerable sense. Fighting fires was what federal agencies did; it made them a stronger presence in the West, where officials were not always welcome, and it also was the most basic human response to the problem of fire. The National Park Service's problem was hardly ideological. Instead, the real difficulty became finding a way to replicate the army's approach without having the resources of the War Department at its disposal.

Sequoia National Park provided a rare testing ground for alternative ideas about fire. As a species, sequoias offered a natural counter to the idea that suppression was a sound management alternative. The big trees had existed for millions of years, and fire was essential to their propagation. When the National Park Service implemented a suppression approach, underbrush no longer burned regularly, increasing the fuel load surrounding the trees. Fires that had once nurtured the sequoias suddenly became capable of damaging them. An important symbol of the national park ideal, the sequoia was peculiarly vulnerable to management efforts that did not take the species' ecological context into account. From the first moment the NPS stopped a fire in the sequoias, well-intentioned service officials inadvertently endangered one of the symbols they and the American public most cherished.

The superintendent of Sequoia National Park, Colonel John R. White, became the service's most vocal proponent of light burning, engaging in a vigorous debate with NPS second-in-command Horace Albright, himself an unabashed proponent of suppression. The two squared off with vigor, both pointing to their experiences as justification for their perspective. Raised in the California desert, Albright was as hard as the climate that produced him. He was not one to brook dissent on such an important topic, and with his vast influence and reputation for punishing opponents, much of the service rallied around him for political as well as ideological reasons. Charles Kraebel at Glacier National Park weighed in, calling light burning an "ogre," after Albright called the practice "unsound and fraught with an enormous amount of danger." Isolated, White persisted, asking for a new look at the idea.[23]

White had watched the aftermath of the disastrous California fires of 1924—which largely missed the national parks—and recognized the flaws in the suppression strategy. He complained in 1926 that suppression practices contributed to the worst fire in the region's history, a 120-square-mile blaze that destroyed 10 square miles of park timber that year. The "forest floor is thick laid with a mass of combustible pine needles, branches, logs, and snags," he insisted in his 1926 annual report, "which makes a conflagration almost impossible to control." White devised his own remedies, national policies to the contrary. In at least one instance, White ordered his men to burn an area—much to their consternation—and with a staff of rangers on hand the first intentionally ignited controlled burn in the park system was a success.[24]

White held a minority perspective. The institutional power of the Forest Service and its role as the lead agency in firefighting gave it primacy, subordinating alternative points of view. In 1923, a California government agency had ruled against light burning. The 1924 Clarke-McNary Act, which allowed federal assistance and grants of aid for fighting forest fires, effectively turned fire protection into the basis for cooperative forestry and federal-state relations, and light burning disappeared from discussions about fire management. Chief Forester William Greeley regarded the Clarke-McNary Act as his finest moment, for it cemented the Forest Service's leadership in the field, gave that agency control of considerable resources for forest and fire management, and assured some measure of federal control over what went on in U.S. forests. The act provided an almost immediate influx of cash. By 1928, the Forest Service had received $1 million for the prevention and suppression of forest fires.[25]

Throughout the 1920s, the National Park Service worked to standardize management, creating not only an administrative infrastructure but also independent specialized divisions. The Washington, D.C., headquarters grew in strength, with Albright as its chief legislative liaison even as he served as superintendent of Yellowstone. Without an intermediate layer of management, superintendents communicated directly with the nation's capital, bringing local problems squarely into the view of NPS leadership. The NPS's only field office, in San Francisco, handled construction and development, adding the beginnings of what became interpretation, the National Park Service's art of explaining the features of the park to visitors, and resource management in the early 1920s. Californians played a prominent role in the development of NPS management philosophies and strategies. The appointment of Ansel F. Hall, another University of California graduate, as the National Park Service's first chief naturalist in 1923 began a process of promoting innovative park-level staff to leadership positions. Naturalists often were assigned responsibility for fire management at the park level, and Hall was set up to be the counterpoint to White.[26]

By the mid-1920s, the National Park Service's response to fire alternated between policy and operational realities. Albright's power and standing guaranteed that the National Park Service would choose his view over White's. Nevertheless, individual parks made strong progress in light burning. Under John White, Sequoia National Park implemented a full-fledged fire regime, initiating what amounted to controlled burns on a number of plots and experimenting with the strategic use of fire throughout Sequoia. White was able to engage in a clear violation of National Park Service policy for two distinct reasons. First, he commanded great respect within the agency, his service dating almost to the founding of the NPS; second, he had the good sense to keep the knowledge of his practices within a small circle, many of whom were at least ambivalent about suppression as a goal.[27]

In contrast, the NPS followed the Forest Service's pattern at Glacier National Park, which by 1926 seemed to be prepared to grapple with fire. In the course of the decade, a combination of factors, including funding, equipment, and technology, had changed this park from a place where fire had been a consistent problem to one where the service had mustered the limited resources it had to fight fire. Officials believed they had a plan to successfully battle it. Even the replacement of Superintendent Walter Payne with Charles Kraebel, a veteran of the Forest Service, reflected a new National Park Service aggressiveness about fire. Kraebel promoted himself as knowledgeable about fire, engaging in debate with well-known authors about the efficacy of light burning.[28] A new era seemed ready to dawn, but whatever confidence developed was shattered in August 1926, when the worst fires to hit the inland Northwest since 1919 began.

By the end of July, fires had created a crisis in park management. Summer weather conditions mirrored those of 1910, with extended high temperatures, little moisture, and significant winds. Blazes started in five separate locations in Glacier, spreading so rapidly that by the first of August, 19,000 acres of timber were burned or in serious danger of burning. Among the sites affected were areas along Fish Creek, on the North Fork of the Flathead River, and in the Blackfeet National Forest, west of the park boundary. The earliest of these originated in the national forest on July 7 and jumped the park boundary five days later. Kraebel responded with the resources at his disposal, but not as vigorously as people in the area had hoped. Some residents expressed their ire. One of them, W. A. Boz, held a personal grudge against Kraebel, and Boz's relationship with Montana senator Thomas J. Walsh, an early nemesis of the service, earned his complaints a serious hearing. In response, the Department of the Interior dispatched Horace M. Albright to Glacier to take charge of firefighting inside the park. On August 3, Albright filed his first report from the blaze. He announced the establishment of eight fire camps with a force of 425 men. As always, the adept Albright demonstrated rapid control of a situation that only days earlier seemed almost beyond hope.[29]

Even as Albright seemed to establish order, a new fire started. George W. Slack, a salvage logger who had been reprimanded for lax operations that had led to fire earlier in the year, continued to operate within Glacier. On July 31, with a high wind blowing, a gasoline tank on one of Slack's trucks exploded, igniting a fire that spread quickly. It reached Lake McDonald, one of the primary features of the park, which was crowded with travelers late in the summer. Albright feared that the fire would ruin the scenery, and it certainly posed a threat to the many visitors in that area of the park.[30]

This particular fire was "intensely hot," Albright noted. The understory, the combustible material that builds up on the ground in the absence of sporadic fire, was particularly thick, mute testimony to the success of localized fire suppression since 1910. "The timber is thick and heavy, and the ground is covered with brush, down timber, and deep humus of pine needles and rotten leaves," Albright observed.[31] Although no one at the time recognized the connection, the Lake McDonald fire graphically illustrated the inherent problem of suppression. Success in fighting all fires thickened the understory and created a more powerful fuel load, guaranteeing a hotter and more destructive fire when an area finally did burn.

Albright recognized the danger of the Lake McDonald fire and mustered all the resources that could be spared. He drafted a hundred men from construction crews, moving them to the west side of Lake McDonald on August 5. In the next twenty-four hours, they dug trenches to the summit of Howe Ridge, blunting the fire's advance. At the same time, high winds—the most powerful that many long-time Montanans recalled ever experiencing—drove the West Huckleberry fire down the north slope of Apgar Mountain until it merged with another branch of the Lake McDonald fire. The meeting of the two fires, Albright told Mather, "compelled considerable readjustment of our fire fighting organization." Digging trenches in front of these fires required the redeployment of a number of the men and took almost four days to accomplish. By August 14, Albright felt he had the majority of the fires under control. Only a little trenching work remained, he reported, and nearly all the fires were contained. Rain would accomplish the rest, which it later did.[32]

The Glacier fires in 1926 were a disaster for the National Park Service. Not only did more than 50,000 acres of timber burn at a cost of almost $230,000 even as the NPS employed 3,583 men to battle the blazes, but the service also found itself on the political defensive. Many elected officials on the northern plains already evinced strong anti–federal government sentiments, a legacy of earlier struggles about federal reservation of land and conflicts with Mather's agency. Attacked by the powerful Senator Walsh, the fire situation challenged the dependability of its park superintendent, an experienced Forest Service man, and even the adroit Albright. A complicated ecological event landed in a distinctly political context. Worse, the service had to fight these fires without even the promise of resources. The National Park Service had only $38,000 in its budget for fire management in the entire system that year.[33]

The 1926 fires galvanized the NPS into a systematic response as the 1910 fires did the Forest Service. While individual parks had always lacked the resources to adequately fight major conflagrations, the 1926 fire season elevated that problem to a national issue. The fires also highlighted NPS vulnerability not only to political leaders—something the NPS well knew after being in existence for a decade—but also to local constituencies and the media. These lessons contributed to a more sophisticated response to fire in general.

The National Park Service turned to chief naturalist Ansel F. Hall, who took on the role of chief NPS forester after the 1926 Glacier fires. The adept Hall had risen quickly in the National Park Service. Hall graduated from the University of California with a degree in forestry in 1917 and served as park naturalist at Yosemite from 1920 until his 1923 promotion to chief naturalist at age twenty-nine. Even before becoming chief naturalist, he had begun an extraordinary career in institutional development, raising private funds for the Yosemite museum and showing the creative leadership that became the hallmark of his career.[34]

Albright had advocated a more specialized NPS management structure, with experts in specific areas of significance. The new approach to fire, the recognition that it was systemwide and endemic, required a general manager on the national level. Still, Hall had many responsibilities in the service, limiting the amount of time he could devote to fire management. Within two years, Hall recognized that fire management was beyond the capability of NPS naturalists. By some accounts, his emphasis on collecting material culture and designing museums helped to prompt his recommendation to Mather in 1928 to create a position called "fire-control expert."[35] This job, different from forestry in some ways, reflected the growing specialization of the NPS as well as Hall's recognition that fire management required a great deal more time and energy than it had received.

The move to find a forestry specialist for the National Park Service accelerated after White reopened the light-burning controversy. In August 1928, in the aftermath of a fire that crossed national forest, national park, and state lands, initial newspaper reports in California claimed that because of White's belief in light burning, the NPS had not responded to the fire with all of its capabilities. In truth, the investigation showed, the service aggressively fought the fire inside and outside Sequoia's boundaries, and it spent more per acre than any of its counterparts. Still, White's beliefs attracted negative attention and seemed to affect his rangers' morale. Albright often fired people for public disagreements with National Park Service policy, but White dated from the earliest days of the service and remained a proven administrator. Rather than level his wrath at a dependable man who had been an important part of the NPS so long, Albright simply assured other agencies that White's emphasis on light burning did not reflect NPS policy.[36]

The creation of the Forest Protection Board in 1927 further isolated White. The 1924 Clarke-McNary Act had created the context for greater federal-state cooperation through the Forest Service, and the board was established to coordinate fire management activities among federal agencies. Other agencies easily bent to the Forest Service's perspective. Not only had that agency made great strides in fire suppression under chief forester William Greeley during the 1920s, but the Forest Service still held the only blank check for firefighting. With his calls for light burning, White made the NPS look unruly and out of touch, threatening the image that Mather and Albright sought for the NPS.

With the controversy raging, the National Park Service looked for someone with a strong background in fire management and the ability to explain and address the different facets of this contentious issue. With California figuring so prominently in the light-burning controversy, service officials focused on finding someone with experience in the Golden State. They settled on John D. Coffman, supervisor of the California National Forest (now the Mendocino National Forest), home to numerous light-burning advocates. An experienced forester and forest manager, he brought a compendium of skills and knowledge that the NPS had not previously possessed.

He had successfully squelched light burning, an attribute that increased his desirability to the service.[37]

Coffman found little to impress him when he became the sole fire control expert in the NPS. As late as 1929, the service's entire fire corps consisted of Coffman, a special fire organization at Glacier National Park that was one more result of the 1926 fire, and a sole fire guard at Sequoia National Park. Although he actively developed larger administrative functions, in particular making the NPS an important presence on the Forest Protection Board, Coffman faced the same problem that had vexed the service since it began. Treated as a local rather than as a national issue, fire commanded limited resources. Congress did not allocate sufficient funds to manage fire in the parks, and the NPS's national leadership persisted in seeing fire as a local problem that episodically flared to higher levels. Although Coffman's hiring was an important symbolic gesture, he alone could not reverse four-decade-old trends. The dollar figures remained ludicrously small: in 1928, the year Coffman was hired, the NPS received $30,000 to manage fire on its 6,133,614 acres. Unless, as White had argued in 1926, the service could secure appropriations "before rather than after needs arise," it seemed unlikely that Coffman's expertise alone could solve the issues that the NPS faced.[38]

Still, Coffman built the beginning of an organizational structure to address fire. In 1928, he produced the first Forest Protection Requirements report, a document necessitated by the National Park Service's membership on the Forest Protection Board and by the desire to have access to the resources of the board. On the ground, his efforts focused on problem parks, Glacier especially. Neither the California parks nor Yellowstone suffered great fires in the late 1920s. Nor did other parks experience severe fire seasons during this generally wet time, providing the NPS's first fire boss with a little breathing room. Coffman could focus on his greatest problem, Glacier, where his special fire force staffed four lookouts on the park's western side, while he considered a fifth lookout on Mount Brown. The Forest Service's Moran ranger station provided training for NPS fire guards and lookouts, another measure of cooperation between the two feuding agencies. In the early summer of 1929, park guards and lookouts detected and suppressed thirty-seven fires, a major accomplishment.[39]

Coffman also introduced fire planning to the National Park Service, a major step forward. He transferred Forest Service planning procedures, which had evolved from regional forester and Forest Service stalwart Coert duBois's classic *Systematic Fire Protection in the California Forests.* The NPS had accepted the ideas in principle before that time, but had not acted on them in any systematic way. The duBois model was a plan for efficiency: an operational strategy, it told how to do better what the service decided to do, not what it ought to do. In this way, policy followed from procedures. The Forest Protection Board mandated the development of fire control plans, and in an effort to show its sincerity about membership in the organization, the National Park Service ordered fire control plans for all its parks.[40]

Glacier became the National Park Service's focal point, for no park in the 1920s provided a better example of the need for fire planning. At the time the service hired Coffman, Glacier had become the NPS's major fire problem, and its issues dominated the thinking of the service's new and enthusiastic fire specialist. In interviews many years later, Coffman recalled the time as exhilarating. He believed that Glacier served as a template for the issues of fire in the national parks and that if he could solve the problem of fire there, he could create a model that would apply throughout the park system.[41]

In the spring of 1929, Coffman and chief ranger F. L. Carter drafted a fire control plan for Glacier National Park, the very first such effort in any national park area. The document outlined an organizational structure, with the park's chief ranger as its fire chief and each district ranger responsible for prevention within his area. All lower-echelon fire crew members reported to the district ranger. A dispatcher at the park headquarters initiated action on all fires and kept track of the disposition of firefighting equipment. The plan dictated reporting requirements, emphasized the need for cooperation with other agencies, and provided instructions for the maintenance of communication sets and other equipment.[42]

Beginning in 1930, Coffman implemented the same fire plan at other parks. By doing this, he created a model of National Park Service fire planning that was derived from the Forest Service but entirely unique to the national parks. Remarkable attention to detail accentuated the gravity of the issues addressed in the plans, and the emphasis on structure reflected both the military character of the early National Park Service and the inherited legacy of the Forest Service's need for close administration of its decentralized agency. The tight organization that the plan demanded spoke volumes about the need for precision and dependability when fire struck. For the better part of a decade, in Coffman's likely biased estimation, the Glacier plan stood as the service's best.[43]

The reality of fire differs greatly from even the most meticulous plans, however, and another serious fire at Glacier seriously challenged Coffman's structure. In August 1929, just a few months after the plan was adopted, a fire broke out in slashings cut by the State Lumber Company on private land ten miles from the park boundary. It began at 4:40 p.m. on Friday, August 16, and was reported almost instantly. Thirty-five men promptly fought the fire, a number that quickly increased to 165, but they could not stop it in the dry conditions and high winds. Wind proved the catalyst for the fire's rapid spread, negating the efforts of men on the fire lines and pushing the fire beyond the trenches they dug. The NPS watched carefully during the first few days, its rangers visiting the fire camps outside Glacier. As late as August 18, most agreed that the fire would not reach the park. That assessment proved too optimistic. "The wind blew hard all Sunday night," Coffman recounted in his summary of the fire, and on August 21, five days after it began, it jumped the park boundary.

A crown fire "of the most destructive type," Superintendent J. Ross Eakin later wrote in his annual report, the Half Moon fire was "beyond human agency to stop." Immediately, the call went out for reinforcements. Coffman arrived at Glacier on August 23 to bring administrative and front-line experience. The NPS hired temporary firefighters but could not stop the blaze. It burned through the park, missing the headquarters building by a half mile, but burning buildings at Apgar and scorching about two miles of shoreline at Lake McDonald. The fire raged for more than two weeks. "Despite desperate efforts to check it," Horace Albright observed, the fire proved its "particular viciousness." Only after a rain in early September did the fire die down enough to allow fire crews from other parks to begin to return to their regular duties. Finally, on September 5, Glacier declared the fire under control. In the final damage assessment, as much as 50,000 acres of timber inside the park had burned and an equal amount of damage outside its boundaries was reported.[44] Fighting the fire cost more than $120,000 and required more than 700 firefighters at the peak.

The analysis of the Half Moon fire represented a new dimension in the NPS's response. Coffman took the lead, authoring an authoritative twenty-page report that included a chronology, the various perspectives, and the most solid analysis of a fire the NPS had yet produced. By the time the blaze began, the National Park Service had successfully suppressed twenty-nine fires inside Glacier and another fourteen beyond its boundaries in 1929 alone. Coffman was "justly proud" of this record and regarded this as proof of the NPS's efficiency. In his view, suppression required "quick detection followed by quick attack with an adequate number of firefighters." The service met that test all year; Coffman believed that the failure of the private timber company to provide rapid response was the factor that allowed it to spread to Glacier.[45]

Coffman strongly defended the service. Understaffed, National Park Service crews concentrated on protecting government property, particularly the park headquarters. This decision drew the ire of private landowners inside Glacier's boundaries. Some even demanded payment for the work they did to protect their own property. Coffman persuasively argued that landowners outside the park provided their own fire protection, often hiring firefighters from the same pools of labor that the National Park Service utilized. He saw no reason that the homesteaders who had settled inside Glacier before the establishment of the park should expect preferential treatment and saw even less justification for their claims of reimbursement. Superintendent Eakin made a convincing argument that the destruction of the park headquarters, a $200,000 investment, would have crippled the National Park Service's firefighting response. "Had Headquarters burned," he insisted, "we would have been practically helpless to combat the fire after the high wind subsided. . . . Had I scattered our forces and lost Headquarters, my position would be untenable."[46]

Even more, the National Park Service learned important lessons about the power of fire. The Half Moon fire had been quickly detected, but even rapid reporting did not allow a more comprehensive response. Instead, Half Moon mirrored the fires of 1910 and 1926 at Glacier, in which people did what they could, but only rain on September 2 dampened the fire and made control possible.

Half Moon illustrated the agency's primary problems: catastrophic fire remained impossible to contain, and its impact negated many of the positive results of planning and preparatory work. Coffman's strategy worked with small fires. But all the preparation in the world meant little when park officials were faced with an enormous and out-of-control fire. Without better technology, more resources, more people, and better planning, the National Park Service could not manage large fires. Even more, fires that began outside park boundaries asked difficult questions of NPS managers. The service had to determine whether it should fight fires beyond its borders as a preemptive strategy.

Even as the National Park Service dealt with the aftermath of the Glacier fire, larger economic changes altered the climate in which it operated. The stock market crash of 1929 hurt funding prospects. Within a year, the federal government slashed expenditures, cut programs, and otherwise sought to staunch the flow of dollars. The National Park Service's popularity with the public stood it in good stead, but the service's role in society seemed frivolous when almost a quarter of the nation's population was out of work. The funding to support programs could not be found.[47]

By this time, the service had accepted suppression and fire exclusion as agency policy. Dissent still existed, but it had been confined, isolated, and explained to other agencies. At the same time, fire exclusion was fantasy. No agency, not even the Forest Service, could fight every major fire, and no one could really do much about major blazes.

Yet the same dilemmas remained, made worse by the Depression of 1929. The National Park Service lacked the resources to fulfill the mandate it had laid out for itself. Unlike the Forest Service, which by the 1930s had a broad enough institutional base to assure more than a tepid response in most circumstances, the NPS had no depths to plumb, nowhere near the fiscal reach of the Forest Service when it sought short-term labor such as casual firefighters. This poverty turned out to be a disguised blessing, although one that the service leadership did not recognize at the time. Because the NPS could not suppress fire with the vigor it wanted, fire and the ecological benefits it brought persisted in many places in the national park system.

By the time Franklin D. Roosevelt was elected president in November 1932, four years after Coffman had been hired, the National Park Service could point to significant improvement in its understanding of how to address fire, but as a result of the Depression, the NPS had not improved its ability to respond. The agency had become a follower rather than a leader.

THREE A Decade of Transformation
The New Deal and Fire Policy

Before the New Deal, the image of fire protection in the National Park Service exceeded the reality of implementation. Without adequate resources, John D. Coffman's meticulous fire plans and the strategies he proposed for the park system remained a far-off ideal. The National Park Service had recognized the problem and created a context for addressing it, but until resources were devoted specifically to fire suppression, controlling and responding to fire competed with everything else that occurred at a national park. The nation's premier national park, Yellowstone, faced the same predicament as every other park area. The resources available were not equal to the task of fire management. Fire protection—the art of recognizing fire and quickly responding with a suppression strategy—became a stand-in for suppression even at the brightest of the service's crown jewels.

Managing fire in national parks posed not only financial but intellectual problems. Before the New Deal, the question was always the availability of resources. The New Deal provided resources in ways that previous managers could not have imagined, but it did not solve the dilemma of how to deploy them. As a result, the national parks benefited in all kinds of ways from New Deal programs. They received work forces that built trails, firebreaks, and even lookout towers. New roads let firefighters reach fires more quickly, and trucks loaded with workers and equipment rushed to meet fires.

But the introduction of all these resources did not advance agency thinking about fire. The NPS continued to follow the Forest Service's lead, adhering almost completely to the objectives of suppression. The enormous power of the Forest Protection Board and Coffman's roots in the Forest Service stifled any counter intellectual energy devoted to thinking about how to manage fire. While the NPS and the national parks benefited tremendously from the New Deal, firefighting advances and the thinking about fire stagnated.

In 1929, when Coffman developed Yellowstone's first fire analysis, he announced that protection was "not normally a very serious [problem]," as long as crews "discovered and controlled promptly" any fires. This set a standard, creating an expectation that all fires would be detected and put out. Coffman found the number of Yellowstone fire lookouts—only two, one on Mount Washburn and the other on Mount Sheridan—inadequate to the task. Neither lookout contained sufficient facilities to allow a spotter to spend the day and night. In most cases, a fire observer made a daily trip from the nearest station to the lookout post.[1] The situation at Yellowstone was better than at most parks, but it was far short of what Coffman and the National Park Service envisioned.

Fires in Yellowstone during the summer of 1931, two years after implementation of Coffman's fire plan there, characterized the limits of protection before the New Deal. Yellowstone had not experienced a serious fire since 1919, and even before 1931, Superintendent Roger Toll worried that the park staff had grown complacent and was not terribly vigilant. Yellowstone simply did not experience serious fires, many insisted. That summer, they were proven wrong. The park faced 112 fires, 25 of which required fire crews. More than 20,000 acres of timber burned before crews contained the outbreaks. No one could claim after 1931 that fire was not a threat at Yellowstone. "The experience of the past summer," Toll drolly observed in his annual report, "has thrown this belief into the discard pile."[2]

From another perspective, Coffman's plan for Yellowstone revealed some success: the fires were managed effectively. In a drought year, with a hot dry summer following a mild winter with light snowfall, Coffman's preparatory system assured rapid response when a fire started and lookouts quickly discovered its existence. Several fires broke out in remote areas in Yellowstone, but the park's fire stations soon sighted them. With hard work from the road maintenance crew and the rangers and deployment of the caches of fire equipment and supplies upon which Coffman had insisted two years earlier, the park was able to keep those remote outbreaks from becoming serious threats. Even the worst fire, an 18,000-acre blaze near Heart Lake, was under control at the end of the second week of response. Especially during a serious fire season, Coffman's foresight seemed justified.[3]

From Superintendent Toll's point of view, the plan's cost to the park was simply more than could be borne. He calculated that the total cost of fighting the fires exceeded $150,000, stripping Yellowstone of a good portion of its road maintenance budget and limiting facilities development. "The entire ranger personnel as far as possible was used on fire protection work,"

Toll noted, at great cost to nearly every other park program. Efforts to control blister rust and insects such as mountain pine beetles were drained of resources.[4] The fires may have been stopped, he averred, but the rest of the programs at the premier national park suffered.

Yellowstone received special attention even during the worst of times. In 1932, in part to compensate for the expenditures that resulted from the 1931 fire, the park obtained $122,780 in emergency reconstruction and firefighting funds and an additional $16,300 dedicated to forest protection and fire prevention, sums so grand that they represented five times the amount allocated for fire in the entire park system just a few years before. The injection of resources prompted greater planning among federal agencies involved in firefighting in the Yellowstone region. In 1932, three important fire conferences took place in the park. The first brought together the heads of all government departments and the protection staff in the vicinity to discuss fire principles and to develop strategies for cooperation. John Coffman also presented a three-day fire training symposium.[5] That year, Yellowstone engaged in more planning than in any previous year.

Yellowstone demonstrated that, although suppression seemed an attainable goal, the service could not implement it and accomplish all the other priorities that NPS managers, Congress, and the public had for the park system. The service's most basic priorities remained land acquisition and facilities development to accommodate an ever-growing stream of visitors. After 1929, NPS director Horace M. Albright turned much of his attention to acquiring eastern parks and the nation's historic sites.[6] When NPS leaders received resources, they still typically devoted them to visitor services or access to national parks. Fire was a recognized threat to those parks, as Coffman's hiring attested, but planning a response and providing the resources to implement it were significantly different steps in a long process.

As the 1932 elections approached, Albright and others in the National Park Service recognized the scope of Coffman's accomplishments and invited him to the annual national park superintendents' conference in Washington, D.C., to brief NPS officials on his efforts. The superintendents' annual meeting was the most important event in the National Park Service's operational year, and Coffman's appearance accentuated the significance of his work.[7] The NPS cadre of chief field officials heard from the man in charge of fire, learning what they might do when the inevitable reached them. A significant problem remained: without resources or a staff dedicated to the purpose, much of Coffman's planning seemed beyond the service's reach.

The New Deal and the resources it furnished changed the context in which the National Park Service operated. Federal largesse permitted the implementation of Coffman's program, a significant fire suppression regime backed by enough work power and resources to inspire confidence in the idea of fire suppression in the national parks. President Franklin D. Roosevelt's constellation of programs to revive the U.S. economy and with it the national spirit transformed the country. Before 1933, federal government officials had only a peripheral role in the daily lives of most Americans.[8] Roosevelt recognized the need for government action, and the New Deal's many programs soon provided work for a distraught nation, primed the pump of the sluggish economy, and inspired hope.

At the same time, the New Deal transformed conservation into labor policy. Under its auspices, conservation programs ranked as highly as capital development ventures; both put large numbers of people to work. The Tennessee Valley Authority, an enormous regional planning program designed to help the impoverished people of Appalachia, provided one component, as did programs to rehabilitate overgrazed Indian reservation lands and the Dust Bowl regions of Kansas, Colorado, New Mexico, Oklahoma, and Texas. Conservation required intensive work power, and no commodity existed in greater abundance in 1930s America.[9]

A few weeks into his presidency, Roosevelt proposed that an army of unemployed people be sent into the rural parts of the nation to perform basic work on federal and state land. They would work in forestry by clearing brush and trees, cutting fire trails, preventing soil erosion, and helping with flood control projects.[10] The creation of a federally funded work force gave land management agencies the opportunity to implement conservation programs that prior to 1933 had simply and completely been beyond their reach. The Civilian Conservation Corps developed plans for work on federal lands, hired hundreds of thousands of young people, and kept them at work in six-month increments called "enrollment periods." Young people flocked to these programs in search of opportunity and stayed as long as they could.[11]

The CCC was a godsend for a struggling nation. It took single people between the ages of eighteen and twenty-five and gave them hard physical work on the forests, parks, and other public lands of the United States. CCC workers were counted among the fortunate during the Depression. The young people lived in barracks and worked six days a week for $30 a month, all but $5 of which was sent home to their families each month. They built roads, trails, firebreaks, structures, and a range of other necessities and amenities on public land. During its nine-year existence, more than 2 million enrollees worked in 198 CCC camps in national park areas and 697 camps in state, county, and municipal parks. The national forests and other public lands contained countless others. Under the bureaus that administered CCC programs—the Emergency Conservation Work program (ECW), the Public Works Administration (PWA), the Works Progress Administration (WPA), and others—crews built more than 1,000 miles of park roads and 249 miles of parkways in national park areas.[12]

The canny NPS director, Horace Albright, recognized in the New Deal an answer to every resource need the National Park Service had. Developing goals that meshed with the New Deal instantly became his primary focus. Despite his long history as a Republican, Albright embraced the new Democratic administration, making friends among the new leadership with dazzling speed. He and Roosevelt's secretary of the interior, the irascible Harold L. Ickes, found much common ground. Ickes had been a visitor in Yellowstone in the early 1920s, and he had heard Albright, then superintendent, deliver an impressive talk. Ickes and Stephen T. Mather, the National Park Service's founding director, had been close, and the secretary, a strong proponent of conservation and national parks, wanted to maintain that relationship with Albright. Not only did Ickes spend an extra hour with Albright at their first meeting, the relationship grew into weekend tours of historical sites in the Washington, D.C., area. A suspicious person by nature, Ickes learned to trust the affable if hard-edged Albright, giving the National Park Service and its director an edge as New Deal programming developed.[13]

Albright cemented this relationship with his characteristic personal touch. On April 9, 1933, the director went on the most famous automobile ride in national park history. After lunch on a trip to Herbert Hoover's old retreat on the Rapidan River in Virginia, Roosevelt told his staff that he wanted Albright in the jump seat of his car. In a discussion as the car rolled along the Rappahannock River, the director made his case for the transfer to the National Park Service of historical parks and national monuments administered by other agencies. Albright and the president talked about other things, not the least of which was the value to the National Park Service of the Civilian Conservation Corps. With Roosevelt's approval and Ickes's beaming support, the director embarked on plans that would transform the service.[14]

Long a champion of fire control, Albright gave Coffman a pivotal role in the organization. Albright intuited one of the ideas that became a hallmark of the New Deal—programming that could be applied across geographical, regional, and even culturally different areas. Fire control, which required a similar deployment of resources almost everywhere it was necessary, easily fit such a model. Immediately after Roosevelt's inauguration, Horace Albright assigned Coffman to develop a report that showed how an emergency forestry and public works program could be implemented. The idea of CCC-like work camps had already been formed, and the ever-astute

Albright recognized that the National Park Service could play a significant role and reap important benefits.[15]

Starting on March 15, 1933, Coffman was "busy night and day" developing his report, which he delivered on March 28. Told to come to the nation's capital, Coffman joined Albright in an early April visit to the White House to learn what the president had in mind. Colonel Louis Howe, FDR's private secretary, introduced the National Park Service duo and a number of Forest Service and U.S. Army officials to Robert Fechner, the man Roosevelt had selected to head the Emergency Conservation Work program. Fechner already had a number of plans, and Coffman discovered he was central to their implementation. "I didn't realize at the time that it was going to be eight-and-a-half months before I saw my family again," Coffman recalled in 1962. "During the remainder of that year, I was the busiest I have ever been in my life."[16]

Under Albright's tutelage, Coffman vaulted to a position of influence and power. Ickes appointed the NPS director as the department's liaison to the ECW program. Albright in turn selected Coffman as his designee to serve in this critical role. The selection affirmed not only the importance of fire to Albright, but the director's faith in Coffman. Albright "requested me to work up a program of emergency forestry and public works that could be carried on by these youth camps that were planned for establishment," Coffman remembered nearly three decades later. This department-wide charge was new to the National Park Service, until that time a secondary bureau in the unwieldy Department of the Interior. It also presaged the largest conservation battle of the decade: Ickes's later attempt to create under his leadership a Department of Conservation that included the entire Department of Agriculture.[17]

The pressure on Coffman was intense. The president set a goal of getting 250,000 people to work by July 1, 1933, and although the number strained the limited administrative structure set up for the purpose, the National Park Service strove to meet the objective. Coffman worked at a torrid pace. In May, the service had places for 12,600 workers in 63 camps within the park system. An additional 70 camps were authorized and being prepared to accept enrollees. By the July 1 deadline, more than 34,000 people were enrolled in 172 Emergency Conservation Work camps within the national park system, and the National Park Service had made plans to accommodate many thousands more.[18]

The impact of New Deal programs on the National Park Service changed its trajectory. The NPS base budget increased from a little under $11 million in 1933 to almost $27 million in 1939. In addition, between 1933 and 1937, public works agencies poured more than $150 million into projects in the national parks. The number of camps in national parks rose from 70 in 1933 to a peak of 115 in 1935, continuing with no fewer than 77 through 1941. In addition, the National Park Service oversaw as many as 475 camps each year in state parks throughout the country. As many as 150,000 enrollees worked in National Park Service programs in the peak years, with more than 6,000 permanent supervisors.[19]

The process transformed the National Park Service's fire control infrastructure. In 1930, the entire park system had only seventeen primary fire lookouts. A decade later, as a result of the New Deal, sixty primary lookouts and fourteen secondary structures offered a far more comprehensive ability to recognize and respond to fire. In 1930, the National Park Service employed twelve lookout observers and sixteen fire guards. By 1939, the numbers had increased to fifty-nine lookout observers, fifty-five fire guards, and six fire dispatchers paid by Clarke-McNary Act funds to supplement park rangers. Many others in the National Park Service and affiliated with it had fire protection as a component of their daily responsibilities. In addition, 754 miles of telephone lines, twenty guard cabins, forty-seven fire equipment storage buildings, 522 miles of roads, 1,767 miles of fire trails, 109 miles of firebreaks, and a range of other improvements enhanced NPS capacity. The National Park Service also developed fire danger rating stations.[20]

Fire prevention became one of the primary responsibilities of CCC camps, altering the tenor of the NPS's response to any kind of blaze. After their arrival at national park areas, many crews began to construct firebreaks, remove deadwood, erect telephone lines for better emergency communication, build lookouts, and engage in other fire protection preparation. In the first year alone, ECW literature claimed, the presence of its workers reduced the amount of national park acreage lost to fire by 1,600 acres.[21] The availability of labor obliterated many of the resource issues that attended fire suppression, giving the National Park Service the ability to apply the tenets that Coffman had advocated since he joined the service in 1928. The contribution of federal relief programs to fire management in the park system was astonishing in its scope: 688,255 work days of firefighting and 837,783 work days of fire suppression in the course of the program. Fire as spectacle, creeping over the edge of Glacier Peak to the delight of tourists, was consistent with the service's image of its obligation to accommodate visitors. Wildfire, which might very well have provided ecological benefit, was not.[22]

Now responsible for fire programs on NPS, Indian Service, Bureau of Mines, and state park lands, Coffman found himself with responsibilities well beyond what he had imagined just a few years before as a forest supervisor in the Forest Service.[23] To reward Coffman, Albright created a Field Education Division in the NPS's Branch of Research and Education and appointed Coffman as division head under the branch director, Dr. Harold C. Bryant. The arrangement gave the National Park Service a fire structure that paralleled the Forest Service's. Coffman initially balked at the appointment. His title in the National Park Service had been "fire control expert," with the lead responsibility in forestry falling to Ansel Hall, the Berkeley-trained forester who served as chief naturalist in the service. Coffman's reticence about the offer was understandable. He and Hall had grown close during the more than five years they worked together, and he did not want to be seen as usurping his friend's position. Albright reassured Coffman that he was not superseding Hall, and Coffman accepted the appointment. Coffman's division became the center of the Department of the Interior's fire management activity, putting a transformed policy into practice.[24]

When Coffman assessed models, the Forest Service still dominated the field of fire management. Well before the beginning of the New Deal, the foresters had made suppression their practical religion. Since an important regional foresters' conference in the nation's capital in 1930, the USFS had settled on goals and contemplated the extension of its reach in fire management. This theme became a major element of the Copeland Report, a 1,677-page behemoth formally titled *A National Plan for American Forestry*, which the Forest Service unveiled just after Roosevelt took office. The tacit question that drove Forest Service policy was simultaneously simple and complex: how far could the service extend its systematic fire protection—in geographic, technical, administrative, and financial terms? By the early 1930s, the NPS enjoyed technical capabilities, but complete exclusion of fire remained too much to ask of the era's technologies. An enormous domain and insufficient resources ensured that USFS officials recognized limits to their control. Instead, that service fashioned different categories for the fire protection status of its lands: critical, marginal, and acceptable. Defining what constituted each category proved far more difficult than creating the structure in which they fit. Prompt and thorough protection or no protection at all were agreed upon as the available options for response.[25]

Even as the National Park Service implemented the USFS ideal of suppression, powerful voices in the Forest Service were challenging that model in the aftermath of the 1934 Selway fires in Montana. Their severity prompted Elers Koch, a prominent forester whose personal history with fire stretched back even before the terrible summer of 1910, to question the agency's approach to fire. Speaking of the northern Rockies and the trails and roads that the Forest Service had cut to aid firefighting, Koch saw a mistake—a destruction of wilderness to no avail. He "firmly believed that if the Forest Service had never expended a dollar in this country, there would be no appreciable difference in the area burned over." Such a bold critique of existing practice from someone of Koch's stature guaranteed a hearing for the new set of ideas. In the Forest Service, only the innovative Robert Marshall, who along with a number of other Forest Service leaders was planning the founding of the Wilderness Society, and the brilliant Aldo Leopold, who had left the Forest Service in 1928 and taken a teaching post at the University of Wisconsin in 1933, supported him. Despite Koch's criticisms, suppression remained the dominant Forest Service model.[26]

The Forest Service's research program emphasized fire protection. The goals it set—in particular, the 10:00 a.m. policy, which specified that all fires be brought under control by 10:00 the morning after their sighting or by each successive morning at 10:00—were only achievable with the resources of the New Deal. Federal programs such as the CCC permitted Forest Service leaders to think about fire in much larger terms than ever before. The programs created led to a level of implementation that no earlier era ever matched, and they gave the foresters greater autonomy. The Forest Service began also to use scientists and other professionals in its research program. With this plethora of resources now at their disposal, foresters expected no less than the complete conquest of fire and with it, nature. In this, they began to treat fire as they did other natural elements such as soil and water.[27]

The National Park Service's aspirations in relation to fire were more modest. Coffman's roots in the Forest Service gave the National Park Service an advantage as it structured its fire and forest planning, but at the same time, what he advocated accentuated the USFS vision of fire management and made suppression the keystone of NPS policy. Leaders divided the park system into different categories. Large western parks, usually in the proximity of national forests, always had been the center of thinking about fire. With the threat of major fires ever present, parks such as Glacier, Sequoia, Yellowstone, and Yosemite implemented more widespread suppression programs than ever before, spurred by CCC camps. In 1933, five camps at Sequoia, five at Yosemite, nine at Glacier, four at Yellowstone, and three at neighboring Grand Teton National Park attested to the importance of fire protection.[28] With direct instructions to use CCC resources to enhance suppression, the western parks finally had the resources they needed to mount extensive programs.

The results of the CCC presence and the increased emphasis on planning were almost immediate. The northern plains experienced the same declining precipitation that had prompted the Dust Bowl to the south, and by the mid-1930s, conditions were drought-like. From 1933 to 1935, Yellowstone experienced acute fire hazards resulting from mild winter temperatures, below-normal precipitation, and the early melting of snow from the warmer-than-usual conditions. Despite a large number of fires, the park suffered little damage. Some interpreted this as the triumph of fire control. The new infrastructure and increased resources prevented a replay of past bad fire seasons, proving Coffman correct and tempering Superintendent Roger Toll's pleas for vigilance. During 1933, more than 800 people worked on fire protection projects that included reforestation; clean-up; and road, trail, bridge, and telephone line construction. In 1934, six CCC camps were organized for fire emergencies, with the workers divided into "flying squadrons" of 50 firefighters, with two additional squads of 40 each. CCC enrollees were attached to ranger stations to act as smoke chasers. "They have been of invaluable aid in this capacity," Toll reported, "and in numerous cases have prevented small fires from increasing to considerable size because of their prompt action, and because of the fact that they were immediately available for fire suppression duty." In 1935, all firefighting was consolidated under the chief ranger, with responsibility for building fires moved from the maintenance division to the ranger division.[29]

The results were stunning, testimony to the ability of adequate resources to make suppression a successful strategy in specific circumstances. In 1933, Yellowstone experienced thirty-seven fires, the largest of which was 850 acres. In 1934, only nine fires were reported before June 30, all minuscule in scope. In 1935, the trend continued: the park faced thirty-five fires, only four of which were Class C blazes of 10 acres or larger.[30] In each of these years, Toll had anticipated a severe fire season. Light winter precipitation and dry spring weather made the prospect of fire extremely daunting, but in each of the first three years of the CCC program, Yellowstone was able to control the fires it faced.

The difference at Glacier National Park was equally dramatic. Suppression worked so well that until a freak fire on October 4, 1934, the park had kept its net fire loss to fewer than 100 acres of timber for the year. A combination of the workers from the park's nine CCC camps and the deployment of smoke chasers and lookouts created a near-perfect suppression regime in a short timeframe. The October 4 fire was after the season typically ended and spread wildly. After a snowfall of eighteen inches, the park released its CCC enrollees. "We were so certain that our fire season was over," Superintendent Eivind T. Scoyen observed in the aftermath, "that we had put away all our fire equipment for the winter." With snow on the ground, rangers began to burn brush, a common fall practice. During one burn on the east side of the park, a "wind of almost hurricane proportions," as Scoyen recounted, spread the flaming material into a crown fire that spread along the north shore of Sherburne Lake. Bringing workers from the CCC camps on the west side of Glacier, Scoyen and his full-time staff quickly brought the fire under control.[31]

Other fires at Glacier National Park also presented challenges, with blazes during August 1936 proving particularly taxing. A fire on the Glacier Wall on August 18 became a crown fire before it was detected. With the efforts of almost 500 workers, it was under control by August 22. A dry thunderstorm on the same day started ten fires in the park and more in the adjacent national forest. More than 200 workers were dispatched to these fires, 125 taken from the Glacier Wall fire. Two days later, the combination of rain and the addition of 1,200 new workers from the Forest Service put an end to that blaze. A fire discovered on August 30, a remnant of the August 18 Glacier Wall fire finding new life, spread. It quickly became three separate fires: one on McDonald Creek, another a few miles north of Granite Park Chalets, and a third near Ahern Pass. The intensity of the fire shocked observers. "I have never seen as complete a burn-out as occurred in Swiftcurrent Valley," Scoyen recorded. "With the exception of a few swampy areas, every green living thing, from rocks on one side of the valley to the other, has been destroyed." The fire leveled all of the park's buildings in the Many Glacier area, three of the chalets across the road from the Many Glacier Hotel, and many of the cabins. The hotel was saved because of the efforts of its employees.[32]

The CCC provided an abundant labor supply, something the park system had never before experienced. Superintendents came to see the New Deal as the solution to their problems. Toll recognized the impact at Yellowstone, as did Scoyen at Glacier, White at Sequoia, Charles Goff Thomson at Yosemite, and many others. Scoyen rated the 1936 fire season as "one of the most dangerous ever experienced" in the northern Rockies. "The entire park organization did a magnificent job during this emergency," he informed director Arno B. Cammerer. "Everyone, no matter in what capacity employed, willingly and cheerfully worked day and night without any complaint whatsoever, to bring the situation under control." Even Howard Hays of the Glacier Park Transport Company, a friend of the park who was sometimes critical of its operations, concurred. "Considering the unprecedented drouth [*sic*] to which the Park has been subject," he told Cammerer, "I feel we have been most fortunate to escape without a much greater loss." When workers cleared underbrush and built roads, trails, and fire lines, and especially when lookouts with communications were staffed, reacting to fire as Coffman insisted gave the national park system a very good chance of mastering all but the most cataclysmic of fires.[33]

One of the greatest coups of the 1920s had been the acquisition of major national parks—Shenandoah, the Great Smoky Mountains, and Mammoth Cave—in the eastern half of the United States. Far more heavily visited as a result of their proximity to so much of the nation's population, eastern parks presented a different set of challenges in responding to fire. The humid nature of the region stood in contrast to the aridity and lack of water so common among the western parks. The combination of visitors and long-standing patterns of local use, some of which included seasonal burning, made such parks vulnerable not only to the carelessness that marked national park visitors across the country, but also to intentional fire setting.[34]

At Mammoth Cave National Park in Kentucky, authorized in 1926 but not established until 1941, the National Park Service faced the problem of fire from a new perspective. Unlike western parks carved from public lands, Mammoth Cave and the other eastern parks had to be bought, parcel by parcel, from private landowners or obtained by cajoling, negotiating, or exchanging with state, county, and local governments. At Mammoth Cave, two associations purchased land for the park throughout the 1930s. Before the formal establishment of Mammoth, the National Park Service took administrative responsibility for the lands that were to be included in the park. This long and often drawn-out process left the service with vast and scattered holdings that were hard to manage and even more difficult to consolidate.

Mammoth Cave had a different pattern of fire than did most of the parks in the system. Fires occurred in its vicinity in winter, with many just outside the park boundaries and even more on private parcels inside the proposed park. NPS officials worried that fires on private land might spread and damage the park, and in any event, they expected eventually to acquire such lands. If the NPS aggressively battled fire, the agency also stood to gain friends in an area where it was resented.[35] "Due to CCC labor we are in a position to suppress numerous fires outside the park," observed NPS representative Robert P. Holland in 1935, "and thereby assist our fire prevention program by making local people fire conscious."[36] In the East at least, the National Park Service served the function of the Forest Service in the West, as the dominant fire response and control agency.

By the end of the decade, the National Park Service had pulled back from this policy. Experience showed that even with the CCC, the task often exceeded the resources available. By 1939, Lawrence F. Cook, head of the Western Region of the NPS Division of Forestry, deemed firefighting efforts outside of park boundaries as beyond the reach of the service.[37] The National Park Service in the West benefited from the fact that its neighbors were primarily federal. With mostly private or state lands around eastern national parks, the NPS found less assistance in implementing a comprehensive fire control program.

The nation's first archaeological national park, Mesa Verde, generally experienced few fires. Colorado's high mesa country offered few opportunities for fire to spread, and at 52,122 acres, the park encompassed a much smaller area than most western parks. In the first twenty-five years of park history, only one fire, the Todd Nine fire in 1926, was considered major. It burned only 20 acres of vegetation, barely reportable by the standards of parks such as Glacier and Sequoia. The park received three CCC camps, which in the summer of 1934 housed 1,300 workers. When two fires started in July, the park responded. The first, the Wild Horse Mesa fire, began on the adjacent Ute Reservation on July 9. CCC workers fought the fire and briefly brought it under control. But the fire broke away, spreading rapidly and eventually subsuming the Wickiup fire, a 286-acre burn that started on July 11. The blazes eventually burned a total of 4,492 acres of timber, 2,229 of which were inside Mesa Verde. More than 1,000 workers battled the blazes, including CCC enrollees, members of the Indian Service, and residents of Mancos, Colorado, and other nearby communities.[38]

The fire created a new consciousness about the threat Mesa Verde faced. Superintendent Ernest P. Leavitt recognized that the park had been fortunate. The fires on Wickiup and Wild Horse mesas did little damage to park facilities; if similar events had occurred on Chapin Mesa, he noted, Mesa Verde's developed areas would have been ruined. Leavitt emphasized the development of a firefighting infrastructure. He wanted lookout towers, truck trails, and fire trails to allow the rapid movement of workers and matériel from headquarters to outlying mesas. The CCC and other New Deal programs provided his solution.[39]

Despite its increasing success, the National Park Service's attempt to eliminate fire became a source of consternation for wildlife scientists within the service. NPS scientists suggested that the Forest Service was clumsy in its approach, its methods heavy-handed and excessive. Under Coffman, some charged, New Deal programs made some national park areas look more like national forests, managed landscapes rather than vestiges of a natural past. In 1935, Adolph Murie, the noted naturalist, challenged existing NPS practice. He believed that clearing a twelve-square-mile area in Glacier National Park as the National Park Service intended was "gross destruction." "Removing natural habitat from a national park," he declared, was tantamount to declaring war on the national park idea. Clearing brush and removing dead trees, denuding roadsides to enhance the visual impact of parks, and otherwise altering existing conditions flew in the face of the wilderness ideal then on the rise. Some scientists vociferously complained about NPS actions, arguing that such human action impeded wildlife patterns, altered terrain, and generally disrupted natural cycles.[40]

Suppression proponents such as Lawrence F. Cook blanched at the accusation that his staff had become "destroyers of the natural." National Park

Service foresters sought to preserve the "natural values" of parks, eliminating excessive fuel loads and maintaining the easy access that promoted fire protection. A protection regime gave nature a better chance of long-term survival, Cook insisted. Without such protection, supporters argued, the National Park Service could not expect to preserve scenic and recreational values or any semblance of native biology.[41]

These perspectives illustrated the gulf between two disparate ideals of national parks. Murie advocated something resembling a fictive pure nature, a physical world that appeared untrammeled to the visitor's eye and satisfied his scientific vision of the concept of natural. Cook argued for a managed scenic landscape, an ideal vista that coincided with the idealized image of national parks, which the National Park Service advanced and the public embraced. In a sense, both fit the definition of nature. Both were managed, albeit in different ways, one by action and the other by the consequences of inaction, and both easily fell within the purview of National Park Service logic and vision. Each pointed to a different kind of fire management future for the service.[42]

Suppression, with CCC manpower, appeared to work. At the same time, an influential countertrend emerged. When George M. Wright, the NPS's first chief of the Wildlife Division, initiated a new plan for wildlife management, the National Park Service had the opportunity to recast its fire response in a manner distinct from that of the Forest Service. Wright and his growing cadre of wildlife biologists never agreed with Coffman's perspective; they liked his policies even less. The biologists believed that leaving dead timber material on the floor of the forest was healthy for the forest and the wildlife that lived in it. *Fauna No.* 1, the first wildlife policy directive the National Park Service issued, advocated preserving the forest as it was, letting natural processes drive any changes in ecology. Coffman's forestry model, extending protection throughout the national parks, attempted to protect them against not only fire, but also insects, fungi, and other threats. Wright's position suggested a dynamic forest, ever changing; Coffman's conceived of a forest frozen in ecological time. The latter remained attractive in no small part as a result of the looming threat of major fire. The National Park Service hired foresters instead of plant biologists or botanists to manage its fire programs, consigning scientists to the narrow realm of plant and wildlife management. Wildlife biologists found themselves alone as advocates of ecological management as the foresters continued to follow USFS practices.[43]

With its tacit value that nature was to be preserved within park boundaries, the National Park Service outwardly embraced the idea of nature preservation even as it developed tourist facilities in national parks and made other significant accommodations for visitors. Fire suppression was not incongruous with the service's vision of nature protection, for it preserved a vision of a pristine, prehuman America. The National Park Service could suppress fire and defend nature with only a modicum of discomfort about the contradictions such a formulation contained.

Although a number of other factors clearly contributed to this positive record, the prevailing thinking about fire management pointed directly at the resources available for suppression. NPS internal information bulletins began in 1934 and became more focused on firefighting and forest conservation after 1935. In 1936, Coffman added the CCC-funded NPS state parks program to his responsibilities. That same year, the Branch of Forestry initiated a review of each park's firefighting program. Specialized training for CCC enrollees became common. The idea that fire could be contained through proactive strategies became such a dominant ideology that when Cook later assessed the decade, he drew stark contrasts. "Prior to 1928," Cook observed, "little training or planning for fire protection had been done. As a result, large acreages were burned. . . . With the advent of the Civilian Conservation Corps in 1933, much more rapid strides have been made in completing the most needed physical improvements for protection."[44] This perspective became the baseline National Park Service view.

The National Park Service's preventive fire protection became more aggressive, necessitating greater levels of organization. Because of the vast number of people working in most national parks and the confusion about what they were supposed to do and how they were supposed to do it, a clear set of guidelines became necessary. Coffman developed principles to guide NPS fire protection.[45] The resulting bulletin offered a scientific approach to managing the ground cover that contributed to fires. Coffman insisted that he removed only "dead vegetative matter from the standpoint of *fire hazard reduction* giving due consideration to the requirements of aesthetics and wildlife." This definition resulted from the increasingly vocal complaints of wildlife biologists and the avid work of CCC enrollees, whom the National Park Service often took to task for not differentiating between dead and living material. Coffman recognized that such work should be overseen by trained foresters, but knew that such specialists were in short supply. Direct administration fell to forest technicians who, Coffman insisted, had to be concerned with "furthering the objectives of wildlife and landscaping."[46]

Coffman designed the instructional bulletin to create a common understanding of obligations and the terminology that defined them. He intended to describe conditions and to establish standards for management that could be applied to fire protection activities. "Debris on the ground is a natural condition in all forests," he wrote in a section entitled "Limitations."

> Unfortunately, fire hazard reduction as an ECW project is too often conceived to mean the complete removal and disposal of all dead standing and down material from large forest areas. . . . Fire hazard reduction often serves as an excuse for intensive forest cleanup which is almost invariably ascribable to and governed by an inherent human tendency to tidy up the woods.[47]

This philosophical observation reflected what had become Coffman's dichotomy, the problem of doing the job so well that it damaged the features the park was meant to preserve.

In the end, such bulletins attested both to the success of the program and to the changes it brought to park ecology. When Coffman reminded his charges that complete removal of dead and downed timber was not a primary objective of clean-ups and that wildlife and landscape values had to be taken into consideration, he asserted the values of the National Park Service over those he brought from the Forest Service. Fire protection was a crucial activity, but even to Coffman, it was not a precondition of National Park Service objectives in the way that it was for the Forest Service. Despite its embrace of the USFS model, the National Park Service vision of fire differed. It no longer even nodded toward the dissenting point of view in favor of light burning, which Superintendent John White advanced. In the same bulletin, Coffman called light burning a "practice [that] cannot be tolerated in the national parks." In response to a suggestion from Yellowstone National Park to let remote and valueless timber burn after a summer in which the park lost more than 25,000 acres of timber to fire, Coffman gave a firm articulation of NPS policy. "I for one do not concur with any such policy for the national parks and monuments," he announced. "There are extremely few areas where any fire starting is not a threat to high values."[48] National parks were to look like nature and to be free of fire. In many ways, accomplishing these ends was a more difficult assignment than simple fire eradication.

CCC enrollees proved less compliant than Coffman hoped, and his message had to be repeated throughout the remainder of the decade. Coffman and others in the National Park Service repeatedly issued rules to govern CCC actions and to affirm National Park Service oversight and responsibility. Even under NPS supervision, the CCC sometimes lacked the subtle touch that Coffman and the National Park Service sought. The crews often cleared indiscriminately, a valid response both to their training and to the

USFS model from which it derived. The CCC drove fire control as much as fire control directed the CCC.

The National Park Service's language during this era contributed to the confusion. Fire plans typically were aggressive in articulating their intent. "The fire control plan recognizes no Sundays, holidays, or 8-hour days or shifts," Grand Teton National Park's 1939 fire plan enunciated. "When a fire is discovered or reported, immediate action is demanded, and control and patrol measures must be continuously applied without interruption until the fire is out." At Grand Canyon, the park's policy reflected similar objectives: "to reach and combat every fire that starts in the park, or that threatens the park, with such speed, skill, strength, and equipment as to confine it to the minimum of acreage burned and damage caused."[49]

Before 1933, fire response had been a matter of quick action by anyone who was available. With New Deal resources, the National Park Service now had trained people and dedicated materials that it could deploy in a strategic fashion. The burgeoning communications networks in the national parks, combined with the many fire lookouts, allowed for a level of planning that extended far beyond the theoretical response of the 1920s. The New Deal changed the nature of fire plans. They became comprehensive documents that described leadership, responsibilities, and strategies while allocating resources and considering contingencies instead of general statements of goals. In some ways, the fire plans were draconian: in an age when cigarettes were ubiquitous, Grand Teton's document forbade smoking during fire season except in prepared camps and designated places. Grand Canyon's plan permitted the park superintendent to draft visitors to help fight fires.[50]

With an infrastructure provided by its access to resources, the NPS facilitated a series of cooperative arrangements with adjoining national forests that extended the cooperative fire protection that had begun in the 1920s. Fire forced agencies into alliances, relationships that became a hallmark of the New Deal. U.S. Forester William Greeley became the primary advocate of cooperative fire protection. Even before he left office in 1928, structures, such as the Forest Protection Board, that supported interagency responses were in place. The New Deal provided a greater degree of centralized control, which affected all kinds of agencies, not just federal land managers. Most national parks had created relationships with other federal land management agencies. Although such agreements had existed since the 1910s, Yellowstone's 1932 agreement with the Absaroka National Forest served as a model. Before the New Deal, the two agencies were both short of resources, and they essentially had agreed to pool what they had. "Overhead will be loaned to adjacent units insofar as practicable," the agreement read, "without endangering the unit loaning the overhead." Both agreed not to charge the other for anything more than expenses, to deputize members of the other agency when necessary, and to share law enforcement obligations.[51]

The preponderance of resources had changed the nature of such agreements, enhancing their significance and pointing toward comprehensive regional planning. A 1936 agreement between Glacier National Park and the Blackfeet Indian Agency created a "mutual purpose in aiding one another in suppressing all fires as soon as possible with whatever means that may be at hand." At Grand Canyon National Park, a 1939 agreement conceded that the boundaries between the two agencies were artificial and that emergency responses could transcend jurisdiction. "Division lines will not be closely drawn," the document attested. "It is to be understood that there shall be no delay by either organization in going to a fire when there is a question as to which side of the boundary the fire is on." The first crew to arrive was expected to provide the initial response, its leader to serve as acting fire boss until the arrival of the lead person from the agency with jurisdiction. At Yellowstone, a new agreement with the Shoshone National Forest in 1938 extended the park's cooperative arrangement in similar ways.[52]

By the end of the decade, the National Park Service had developed a clear and distinct set of strategies for addressing fires. Service officials relied on leadership at the park level to emphasize the importance of fire response, argued vociferously for careful assessment of fire experiences and for continuous updating of fire protection planning, collected data about the sources and causes of fires, and recognized the value of frequent training for everyone involved in the fire protection system. By 1939, the National Park Service anticipated the end of the CCC. Superintendents were admonished to develop new sources of firefighters in local communities and beyond. "The Service has an enviable position among agencies responsible for fire protection in that practically all the users of the parks are contacted directly by protection personnel," observed Lawrence Cook, a bit optimistically. "We have a wonderful opportunity to advance fire protection not only for our own areas but also in the general field of fire prevention."[53] The National Park Service not only saw its experiences with fire as valuable, it also believed that its educational mission could be used to support the goals of suppression.

Cook further recognized that completely preventing human-induced fires was impossible. "The Service, perhaps, cannot expect 100 percent elimination of man-caused fires despite all that we can do, although our efforts should be pointed in that direction," he summarized. "Any park can well be more proud of a record of reduction of preventable man-caused fires than in a reduction of area burned."[54] This differentiation attested to the lessons that the National Park Service had learned. Prevention was good, but control was essential. More than any other idea, this subtle shift enunciated the differences between the National Park Service and the Forest Service. The primary threat to the parks remained the actions of people.

The initial hostilities marking the start of World War II changed the climate in which National Park Service fire management took place. From 1941 to1945, the war took men and matériel away from civilian purposes. The service's budget was cut in half in the aftermath of the attack on Pearl Harbor. Between December 1941 and the end of the fiscal year in June 1942, the National Park Service lost almost 25 percent of its permanent workforce. A year later, the number had fallen again, from 4,510 workers in 1942 to 1,974 at the end of June 1943. It dropped to 1,577 by 1945. The CCC was disbanded in 1942; most of the young men who worked in it went into the military. The situation became so dire that the service relied on camps of conscientious objectors to open and maintain trails for visitor use and fire protection in Glacier National Park. The National Park Service moved from its Washington, D.C., headquarters to Chicago and slipped into an inconsequential role as the war effort demanded more and more of the nation's resources.[55]

Fire presented a different kind of threat in wartime. "To the normal problem of fire protection," NPS director Newton B. Drury wrote in his 1943 annual report, "an acute threat of sabotage and enemy incendiarism was added." Areas of extreme fire hazard within 300 miles of any coast were included in the fire protection allocations of national defense agencies.[56] The war also increased military use of the parks, providing a faint echo of the army's earlier involvement in national parks. Instead of protection, their purpose now was largely recreation, as Drury noted in 1943 in his famous plea to maintain protection of the national parks. National parks functioned as emblems of Americanism, he argued, embodying ideals for which the nation fought. This symbolic role elevated the national parks' significance and made their protection even more essential. "Their proper protection in wartime is a responsibility of the first magnitude," Drury insisted in his annual report.[57]

Although fire management suffered during World War II, the consequences were not as dire as anticipated. The war may have taken resources from the national park system, but it simultaneously impeded opportunities to travel. A dramatic decline in visitation, in no small part the result of gasoline and tire rationing, meant that a primary cause of park fires—careless people—was in equally short supply. Even the influx of soldiers and sailors did not counteract the decline in visitation. In one astonishing example of the shift, the 1941 Clarke-McNary Act report for Arizona showed seventeen national park areas with more than 1 million acres of forest and no reportable fires. The list included Boulder Dam National Recreation Area, Grand Canyon National Park, Organ Pipe Cactus National Monument, and smaller areas. Fire disappeared as the number of visitors diminished, confirming something that park officials had noted all the way back to the era of the cavalry in the nineteenth century: ongoing fire stemmed from human action with uniform consistency. In 1943, only 308 fires were reported in the entire park system, 23 percent below the average of the previous decade.[58] A number of those were spotted by wartime volunteer lookouts, some of them women. In essence, the NPS receded from the vision of the 1930s, that it would fight fire anywhere it found it, and returned to an earlier vision of battling only proximate fire.

Still, the service's suppression ideology remained constant. Federal land management agencies followed national trends, recruiting older men and some women to replace those who went off to war. The NPS even considered using Japanese internees as firefighters, but found insufficient numbers of men, because most of the people in the age group for firefighting had already volunteered for the war. For the most part, the replacements lacked the "experience and training desirable for most of these positions." The National Park Service and the Forest Service shared workers as well. The "excellent fire programs" that Region II regional director Lawrence C. Merriam observed at Yellowstone and other parks served as the best form of training. They paid "big dividends," Merriam noted, providing an essential component of parks' response.[59]

By the time the war ended in 1945, the CCC and its resources were a distant memory. "For the first time since 1932," Region IV director Owen A. "Tommy" Tomlinson informed his superintendents in 1946, "the agencies handling forest fire protection will not be able to call upon organized mass labor such as the CCC [and] the armed forces."[60] This was a new reality, he instructed his charges, a return to the early days of suppression. Tomlinson asked that his staff do more with less. Suppression remained a powerful intellectual model of response to fire for the National Park Service, but the realities shifted back toward an earlier time.

The New Deal and its resources changed the National Park Service in many ways, and the ability to address fire reflected a prominent improvement in the NPS's ability to fulfill its functions. Suppression had been an ideal; the CCC and its work power, the millions of dollars from public works programs, and the addition of fire specialists to the National Park Service had combined to inspire confidence in its model. With enough resources, suppression worked. Parks faced and fought fires and were able to minimize their impact, except in the most dramatic of circumstances. Even if cataclysmic fire remained beyond the service's reach, day-to-day fires and most extraordinary fires could be controlled without terrific damage. As a result of the New Deal, fire seemed to become one more natural force that human ingenuity could subdue.

This vision, of an orderly intact nature managed by humans, reflected the stance of the foresters in the National Park Service. It also coincided with the views of landscape architects, who remained the driving force in the service. This neatly manicured version of nature contrasted with the messier ideal held by wildlife biologists and other scientists. The foresters' vision held sway even as the signs grew that the success of suppression was only temporary.

FOUR

Ecology and the Limits of Suppression in the Postwar Era

In 1948, Yosemite experienced its first major fire in almost a decade. The Rancheria Mountain fire began on September 9, discovered by a three-man trail crew at about 1 p.m. They had no communication equipment and were more than eight miles from the nearest telephone. The crew assumed that lookouts would spot the fire and notify park managers, so they battled it as best they could, but lookouts did not see the fire until the next day. Beginning in Tuolumne River Canyon about two miles downstream from Pate Canyon, a section that was almost 4,000 feet deep and three and a half miles wide from rim to rim, the prevailing pattern of wind drove the fire upslope, keeping it from rising high enough to be seen by lookouts. Before it was brought under control twelve days later, the Rancheria Mountain fire burned more than 11,840 acres of timber.[1]

The blaze taxed the park's available work power. Late in the fire season, it required a response from a quickly assembled force. Park staff already had been reduced to 10 temporary rangers, 1 ranger naturalist, 5 fire control aides, 47 blister rust workers, and 70 maintenance workers. With the small permanent staff, this comprised the park's entire workforce. The National Park Service was able to marshal a combination of seasonal and permanent staff members to fight the fire. Eighteen rangers, 3 naturalists, 3 fire control aides, 43 blister rust workers, and 85 members of the maintenance crew staffed the fire lines. The Forest Service recruited 55 more casual firefighters in Stockton and brought them to the park. The Yosemite Park Company, the Curry Company, the state fish hatchery, the U.S. Post Office, the city of San Francisco, and the Davis Lumber Company together provided another 153 workers to battle the blaze. On September 16, as the fire began its second week, 220 soldiers from Fort Ord, California, arrived to assist. With the arrival of the military, enough work power had been accumulated to bring the blaze under control.[2]

While the work power could be deemed adequate, Yosemite had weak points in its fire protection strategy. The lack of communication equipment loomed large. The telephone line that stretched from Harden Lake to Pate Valley to Benson Lake was perfectly positioned to report the fire, but the line was out of commission. In 1942, maintenance on the line had stopped as a result of a lack of funds and work power, a direct consequence of the dismantling of the Civilian Conservation Corps. Had the line been operational, news of the fire would have reached headquarters two days earlier, significantly accelerating the response. Radio transmission is irregular in the steep valleys of the Yosemite country, where topography and atmospheric conditions make signals undependable. Insufficient scouting on the fire line contributed to an overall lack of knowledge of the scope and size of the fire, and the perennial need for work power periodically left crucial gaps in the park's firefighting capability.[3] Yosemite's fire protection system needed an upgrade, but the models for change were not developed.

In the postwar era, suppression and a growing interest in the use of fire as a management tool collided, highlighted by new experiences within the park system. Throughout the park system, new ideas came into play, and in the growing and perennially underfunded system, managers at the grassroots level began to experiment in new ways. There simply was not enough oversight or momentum to deter them. Even more, the experience of parks in the eastern half of the nation demanded a different set of precepts for management. The result was wholesale experimentation with fire in some eastern parks even as the Sierra Nevada parks, Yellowstone, and Glacier continued with the old suppression regime. But the tension between the perspectives was palpable, and something had to give.

The end of World War II inaugurated a new era for the National Park Service, one of unexpected growth and precipitous change. The enormous increase in the number of travelers and their desire to experience the national parks pulled the NPS from its historic moorings and compelled the service to envision new ways to manage its holdings. Postwar Americans appreciated their national parks in ways that their parents could not. Beneficiaries of a revolution in expectations, access, and affluence, Americans visited the parks to see their country, to feel its power, and ultimately to understand themselves.

Such behavior was part and parcel of a larger transformation of U.S. culture and society. Almost everything about the nation—from race relations to recreation—changed as a result of the victory over fascism. A new enthusiasm swept the nation, a sense that everything was possible and life would get better for everyone. Indeed, evidence to support this idea appeared everywhere. The position of African Americans changed dramatically, first in symbolic ways such as the integration of major league baseball in 1947, then in more substantive moves such as the integration of the military in 1948, and later in the law with cases such as the Supreme Court's 1954 *Brown v. Board of Education* decision, which declared "separate but equal" education facilities to be illegal. Federal home loan legislation made homeowners out of renters; veterans went to college on the G.I. Bill. Americans bought houses by the thousands in new suburbs such as Levittown on Long Island, New York, achieving the new American dream.[4]

The postwar era encouraged greater appreciation of American nature by a broader cross-section of the public. Within a few years of the war's end, many Americans enjoyed greater disposable income and paid vacation time, and millions used them to see their national parks.[5] The NPS could not keep pace. Not only were available campsites scarce, but the existing campgrounds were covered with uncollected garbage, debris from timber illegally cut for firewood, and other eyesores. Superintendent John White at Sequoia offered a poignant observation of the conditions that ensued. "In the national parks, we have always been in the position of engineers compelled to dam a stream without opportunity to divert the flood waters," he observed in his 1947 annual report. "It looks as though we must hope for another depression to help the National Parks. Despite our best efforts, our public camps are run down, our scenic spots improperly protected, our park buildings and all facilities inadequately maintained, and the public neither protected nor advised, nor educated." Others soon echoed White. In response to what he regarded as a landscape destroyed, noted author and iconoclast Bernard DeVoto recommended closing the national parks if they could not be better managed. The rapid increase in automobile ownership and the driving vacation had strained the limits of the park system.[6]

Automobile tourism typically took place between Memorial Day in late May and Labor Day in early September, the classic boundaries of summer, when children were out of school, days were long and warm, and families could spend time together. Two days before Memorial Day weekend, tourist camps and motels sat vacant; two days after Labor Day, the cacophony subsided and they returned to silence. Auto usage created clear patterns of travel and behavior. Tourists went everywhere and anywhere; purchased enormous quantities of food, gasoline, and other staples; filled motels and hotels; and generally kept moving, staying only an insignificant length of time in all but one or two of their stops.

These new patterns of vacation travel almost perfectly coincided with fire season in western parks. The NPS encountered more visitors in more places at a time when the resources to manage them remained constant or even diminished. In fire management, the situation dictated that the service would do its best to react to fires, preserving the bulk of its limited resources for crisis situations. For people such as NPS fire guru John D. Coffman, with two decades in NPS fire management before 1950 and aspirations for a comprehensive system of preparation, this reality was a severe disappointment.[7] From the aggressive suppression-based posture of the New Deal, the NPS returned to a pattern of making do in its battle against fire.

At the same time, the National Park Service tacitly began to unshackle itself from the Forest Service model that had driven NPS policy. The Forest Service maintained strong fealty to the tenets of suppression. Its sponsorship of fire science culminated in the National Fire Danger Rating System, the establishment of three USFS laboratories to study fire, the recruitment of new labor such as the Southwest Forest Fire Fighters, the upgrading of fire crews, and especially the transfer of surplus military equipment to civilian fire protection. The National Park Service found it could neither keep pace nor embrace the objective with the same wholehearted enthusiasm. In an era of mechanization, the Forest Service focused on fire protection and suppression. The National Park Service looked elsewhere, investing in the prevention of fires through the education and supervision of its visitors.[8]

During the immediate postwar era, important changes in the NPS response to fire began. At Yellowstone National Park, park engineer Aubrey Haines completed a "Fire Lookout Evaluation Study" in 1946. Following the Forest Service's model, he researched fire records from the previous decade and pinpointed ongoing trouble spots within the park. Haines recommended a fire danger map for Yellowstone, a base document to enhance park managers' ability to detect fire and respond to it. Haines's work was replicated at other parks. The first NPS Interregional Fire Control Training Conference convened at the Grand Canyon in 1949, bringing together fire personnel from throughout the West. Smoke jumpers, frontline firefighters dropped as shock troops to control a fire at its inception, were introduced at Glacier in 1946 and at Yellowstone in 1951.

Smoke jumpers were simultaneously valuable and glamorous. The idea had begun in the Forest Service with David Godwin, an innovative leader and an advocate of technologically sophisticated responses to fire. Under the Aerial Fire Control Project, a direct response to the 10:00 a.m. policy, Godwin attempted to drop chemical retardant on fires from the air in the late 1930s. Soon he replaced the chemicals with people, dropping firefighters by parachute in an effort to attain immediate fire control. Smoke jumpers touched a nerve in the American public; like Pony Express riders, they faced the unknown with courage and aplomb. They heightened the already powerful image of firefighters.[9]

By the time the first Yellowstone unit was formed, smoke jumpers were lionized figures in the battle against fire. Thirteen had died in the Mann Gulch fire in 1949, their deaths later emblazoned on the national stage by a 1952 movie, *Red Skies of Montana*, and later by the 1992 publication of Norman Mclean's *Young Men and Fire*.[10] The desire for smoke-jumping units stemmed as much from the dramatic image they projected as from their real utility as tools to fight fires. At Yellowstone, Edmund Rogers recognized that the NPS had to consistently grapple with peer agencies for preeminence in the public imagination. It did not serve the NPS to have its premier park perceived as lacking any service provided by a mere national forest adjacent to its boundary, even though smoke-jumper units were expensive and arguably of only marginal advantage in fighting fire.

The Yellowstone unit resulted from a 1949–1950 Fire Review and Fire Control Replanning Study. In 1949, Yellowstone experienced a difficult fire year. Five large fires dogged the park. Yellowstone needed a full-time fire dispatcher when fire conditions threatened a major conflagration. The report suggested reassigning an assistant chief ranger to the task, but this was a mere stopgap solution. Emergency lookouts were planned, with extra workers to fill openings in the schedule. The park's single greatest need, NPS Forester Maynard B. Barrows wrote, was "the employment of a plane for aerial detection and transportation of smokejumpers based at West Yellowstone for one month of the year."[11]

Rogers built the second smoke-jumping unit in the NPS. Agency policy dictated that an aerial firefighting response was the most effective and the least intrusive, compared to truck trail construction and other development strategies. Beginning in 1951, the new NPS smoke jumpers were trained by the Forest Service at Missoula, Montana, and transferred to the National Park Service payroll during the fire season. At its inception, the Yellowstone unit consisted of only a five-man crew, but it represented a significant upgrade in response time over the previous Missoula-based USFS unit. Rogers wanted to have his smoke jumpers based at West Yellowstone for the most dangerous part of the fire season, typically from mid-July to mid-September, a longer period than Barrows anticipated. The park leased a hangar at the West Yellowstone Airfield and let bids for a plane.[12]

Ultimately, smoke jumpers were valuable primarily as symbols; firefighting had become an institutional operation, and technological advances dwarfed even the most heroic exploits of any firefighter. Larger trucks delivered more workers to fires, and pumpers gave greater capacity to deliver water where needed. Airplanes served as reconnaissance, spotting fires well in advance of the norms of the previous decade. New kinds of flame retardant were introduced, as were other chemical solutions to fire. The result was a better system to fight fire; smoke jumpers served as the Pony Express, the vanguard that symbolized the technological transformation taking place.[13]

But smoke jumpers were ultimately a Band-Aid on a much larger problem. The coherence of the New Deal era dissipated in the changed postwar climate, and suppression became a reflex rather than a strategy. The postwar NPS placed a premium on devising new fire plans that fit the changing realities of the national park system. The 1949 plan reinforced the 10 a.m. policy that the National Park Service had followed since the 1930s, sought to confine all fires to the smallest possible area, and rearticulated the objective of eliminating all human-induced fires, which comprised the overwhelming majority at the park. In 1948, seventeen of the twenty fires in Yosemite were the result of human action. Clearly, a combination of prevention and education could eliminate much of the fire problem.[14]

The 1949 Yosemite plan epitomized the NPS vision of the response to fire. In nearly everything the service had written since Coffman arrived in the late 1920s, fire was the enemy and the park's job was to put it out as soon as possible. The National Park Service's greatest problem was the behavior of its visitors, who started fires by accident and with intent, leaving park staff scurrying to respond. Detection was a crucial piece of park and service strategy; once fires were discovered, the weight of the intricate fire suppression network could be placed upon them. When they were not or when communications failed, as in the case of the Rancheria Mountain fire, the fire grew out of proportion to its genesis and became a significant problem. In the 1950s, suppression reigned supreme at Yosemite and throughout the western national parks.

If challenged on its strategy, the NPS could point to the success of its suppression programs. Education loomed large: beginning in 1944, the Forest Service had introduced the "remember, only you can prevent forest fires" campaign. By 1953, the National Park Service achieved a reduction of almost 50 percent in acres burned, from an average of almost 27,000 acres per annum between 1947 and 1952 to a new recorded low of 14,833 acres in 1953. The trend continued in 1954, with a 4 percent decrease in human-caused fires and a light year for lightning strikes. At the same time, 1953 was among the NPS's most difficult years for firefighting. Severe lightning fires in remote areas of Yellowstone and Yosemite, combined with drought-like conditions in the Southwest and in California, made vigilance an even more prized commodity.[15]

The decrease in human-induced fires in comparison to lightning fires changed one significant dimension of the NPS mode: fire could no longer be explained simply as a people problem. The National Park Service had to confront nature as nature, not as human behavior gone awry. Lightning fires had been consistent, although often beyond the reach of park capabilities. Yet with the success of suppression, lightning fires flourished. Human-induced burning had competed with lightning for fuels, burning areas that might otherwise have combusted naturally. As suppression succeeded, fuel loads increased dramatically, and on a small scale, lightning burned more of that fuel. Better detection, particularly by aircraft, brought more of these fires to the attention of the NPS. With new technology, the National Park Service found more fires that would have gone out on their own if no one had seen them.[16] The NPS seemed to have traded one kind of fire problem for another.

As visitation increased in the 1950s, the strain on fire protection resources grew. Even as education diminished the number of human-induced fires, the immensity of the task of stopping fire stunned NPS officials. By 1954, visitation increased to a record 47,833,913. In 1955, more than 50 million visitors descended on the national parks; in that year education and prevention decreased the number of human-induced fires from 247 to 173 in the entire system.[17] Lawrence Cook fairly crowed about the accomplishment.

Along with the positive news, a sense of strain simultaneously permeated the annual fire reports. While suppression worked, it continued to attain its goals at the expense of other operational areas at each national park. When they looked at longer trends, managers saw the average number of fires per annum gradually increasing, from 356 between 1945 and 1950 to 362 between 1950 and 1955, which pointed to limits in what could be achieved through educating the public. Fire tabulators were shocked that, in 1956, 422 fires were reported, a much larger number than either of the previous clusters. With the decrease in human-induced fires, the only possible cause could be lightning fires. Worse, the continuing increase in visitors suggested that the number of human-induced fires would also increase.[18] Leaders could not be blamed if, under the weight of dramatic increases in visitors, with the increase in lightning fires, and without a concomitant injection of resources, they feared for their future and pined for another way to address the annual crises that fire caused.

A different vision of the role of fire and the service's response to it existed, but it came from a region of the country that the National Park Service had not associated with forestry. Everglades National Park offered a counter to the suppression model, but the NPS struggled to integrate this park with its vision of fire control.[19] Codified in the NPS *Fire Control Handbook*, a collection of strategies and approaches for administering, finding, and fighting fire, the NPS experience with fire was western in character, shaped in the mountains of California and the inland Northwest at the major fire parks—Glacier, Yellowstone, Sequoia, and others. Despite significant differences in their conditions, they were of a piece. The Everglades represented something different, a low-elevation, wet park that, biologist Daniel Beard, who became the park's first superintendent in 1947, remarked, "burns off twice a year."[20] The Everglades became a counterweight to the National Park Service's generalizations about fire, the place that disproved existing theories of fire management and offered a new approach.

Although authorized by Congress in 1934, the Everglades was not formally designated a national park until December 6, 1947. During the intervening years, the National Park Service acquired land for the park, closely following the parameters that Congress had authorized. This acquisition process paid little attention to questions of management, and only when the park was dedicated did the NPS began to understand what it had obtained. Spread out over much of southern Florida, the new park presented incredible opportunities and even greater challenges.

Within a very few years, the National Park Service recognized that the Everglades, among the first parks established for biological purposes rather than for its monumental scenery, did not fit the model that the NPS had developed for its western parks. According to its organic legislation, the Everglades was established to create a "wilderness, [where] no development . . . or plan for the entertainment of visitors shall be undertaken which will interfere with the preservation intact of the unique flora and fauna of the essential primitive natural conditions." This powerful legislative mandate simultaneously differed from and challenged the park management standards of the era.[21]

In a different climate and environment, the Florida park experienced endemic fire, but observers believed that fire in the Everglades had a salubrious effect. "Within a few weeks after fire, the glades are green with sawgrass shoots, and the pinelands full of flowering herbs and new grasses," observed William B. Robertson. Hired as a fire technician, he gravitated to research and conducted a study of fire at the park in 1953. "Even the scars of burned-out hammocks are soon hidden by rank growth of fireweed shrubs and vines," he said.[22] Robertson's tone reflected the disconcerting nature of what he saw. Fire in the Everglades had a different impact and even a different function than it did in the western parks. It was undeniable, ever present, more acceptable to the surrounding natural communities, and more complicated by the region's peculiar hydrology and biology.

By the time the National Park Service arrived in southern Florida, the patterns of fire in the Everglades had been long established. The region experienced frequent and widespread lightning fires that typically occurred during the May to October wet season and did little damage. Such fires played a crucial role in maintaining many plant complexes. From such fires, observers after 1947 inferred that fire in general had little impact on the park, an assumption that belied a far more complicated reality. Fires in drier periods of the year, November to April, often caused great damage, inducing plant succession by destroying the root systems of even fire-resistant plants. Such fire even consumed dried-out organic soils.[23]

Human fire had an even more pronounced impact. Native peoples had clearly used fire to modify their environment, as they did throughout the Americas and indeed the world. When Euro-American settlers came to the region, their efforts "beggared" their predecessors, Robertson noted. "The frequency of man-made fires probably increased sharply as whites replaced aborigines." In an effort to improve their agricultural prospects, Euro-American settlers began to drain the Everglades. The lowered water levels that resulted increased both the frequency and severity of human-induced fires. An arterial canal system begun in 1905 and the completion of a dike at Lake Okeechobee in 1935 exacerbated existing problems. As drainage became more effective, fires increased in severity. The newer desiccation, especially of the lower glades, extended the fire season by months, pressuring the nascent fire response mechanisms of the NPS. When fires occurred, their intensity led to greater destruction of hardwood forest vegetation and organic soils. Robertson's characterization of an "imposing picture of fire occurrence" as a result of the Euro-American presence provided a strong rationale for aggressive NPS action.[24]

By the early 1950s, the National Park Service had seen enough fire at the Everglades to recognize that conditions there challenged its assumptions about fire control. Its first crisis came in 1950, when three large fires, called Tamiami fire No. 3, Long Pine Key fire No. 3, and Mowry fire, required simultaneous suppression. The park lacked the resources to fight all three at the same time. The "fire emergency merely spotlighted this fact," Dan Beard wrote in the aftermath of a critique held at park headquarters. Beard believed that the park showed strength in "the spirit of the men and women (permanent, seasonal, and temporary) who tackled the undertaking." Despite their valiant efforts, he said, their attempts were inadequate. Although the park could take pride in suppressing three fires, "each of which experienced fire fighters and 'glade cats' said nobody could put out," Beard insisted, "I am determined that the park staff will not be called upon to undergo any repeat performances."[25]

These fires clearly showed that the park's technical capabilities were inadequate. The combination of information and mechanization that characterized the era had not yet reached the Everglades, and the park had not yet developed a fire management plan. Park maps were old and outdated, without roads, trails, and in some cases, terrain and plant distribution. Firefighters started with inaccurate information about the geographic features they encountered and the fuel types and loads they battled. The park's communication systems also fell short. Dependent on walkie-talkies as a result of the absence of phone lines, the park did not own enough radio sets to assure constant communication, and what radios they had lacked sufficient range. Dispatchers lacked training for fire management, park vehicles did not possess sirens, and the park's safety mechanisms were undeveloped. Contingency plans for hiring temporary firefighters did not exist; no one had thought to develop a ready supply of potential workers before fire season. From an infrastructural perspective, the park fell far short in almost every respect, an endemic problem especially for eastern national parks during the early 1950s.[26]

Robertson recognized the combination of employee strain and inadequate firefighting resources as a severe obstacle to managing fires. Five years of firefighting had "absorbed much of the productive energy" of the Everglades staff, he noted, but the results "inspired no feeling more robust than a very reserved optimism." The problems seemed beyond the reach of park staff, a sentiment with which Beard concurred in the aftermath of the 1950 fires. Despite learning a great deal about the park and its fires and developing and implementing a comprehensive firefighting program, the NPS could do little about the real problem—the diminishing amount of water available in the Everglades. The U.S. Army Corps of Engineers planned an enormous flood control project for central and southern Florida. The NPS viewed this project with considerable trepidation. Less water than the already diminished supply was not even conceivable from the NPS view. Without more water, "the best efforts of fire detection and suppression," Beard insisted, "are likely to provide only local victories in a lost war."[27] This tacit admission of the impossibility of implementing policy was a first in the National Park Service and led to the embrace of a different vision of fire in the Everglades.

A move to formalize the use of prescribed fire in Everglades National Park ensued. The park's specific conditions made suppression a dangerous strategy. Officials had long recognized that suppressing fire in rockland areas led to the rapid domination of the landscape by hammock vegetation. Early NPS suppression efforts had succeeded, allowing broad-leafed hammock vegetation to spread, especially along rock pinnacles, along northerly facing sections, and in the wettest areas. In some places, the plants reached twenty feet in height. In all areas, they created a dense understory, the predicate of a "calamitous" fire, Daniel Beard wrote in 1956, "perhaps killing pine as well as understory."[28]

Beard quickly turned into an advocate of the use of fire. In 1956, he argued that its absence promoted the expansion of hardwoods, which in turn would lead to the extinction of the southern Florida slash pine and other pineland plants. Committed to protecting Long Pine Key and other park areas as pineland Beard regarded fire suppression as the chief obstacle to his natural resources management goals. Beard's memorandum catalyzed support for controlled burning at the Everglades. Regional director Elbert Cox sought NPS director Conrad L. Wirth's approval for this controversial plan. Wirth authorized a specific management plan for the project, with the caveat that he see the plan and approve it before it was introduced and that the conservation community be given a look as well. In June 1957, a completed plan reached the Washington office of the NPS, to which the service returned after the war. After much deliberation and a thorough review of the differences between the Everglades and the rest of the park system, in October 1957 Conrad Wirth approved the first controlled burning plan within the national park system in more than thirty years.[29]

The plan was elaborate and persuasive, showing the impact of almost a decade of research in southern Florida. The initial proposal called for one round of burning, with careful evaluation before any additional burns took place. It outlined eleven burning blocks, lettered A through K, with initial plans to fire nine of the eleven. Blocks D and K, which later were redefined as blocks K through Y, were to be left for the future. NPS managers planned to burn backfires into the wind, only lighting headfires sufficient to immolate advanced hardwood succession. The park built twenty miles of roads on Long Pine Key to accommodate the fire plan. By the spring of 1958, the managers of the Everglades stood ready to implement their plan.[30]

When park rangers lit fires in Block B on Long Pine Key on April 21, 1958, they inaugurated the first long-term prescribed fire plan in the national park system. Between 1958 and 1973, 49 prescribed fires were set in Blocks A through Y. Fifty-two more were begun between 1973 and 1979, comprising the vast majority of NPS prescribed burning in that era. Burning during the summer was less frequent; only 13 of the 101 fires in this era were started between June and September. The period between October and January became the favored season. In 1976, time-of-year restrictions were finally removed.[31]

The Everglades presented a fire scheme unfamiliar to the National Park Service. The first generation of fire managers had all learned their skills in the West, influenced by the Forest Service and the legacy of the 1910 fires. The Everglades was different. The NPS had no experience with local residents who said: "This country has always burned and always will, and anyway fires don't hurt anything here," as Robertson was told repeatedly. Yet, officials could identify a pattern that threatened the NPS's long-term ability to protect park values. As the Everglades became drier, sustaining the ecological and scenic status quo became progressively more resource intensive and expensive. The park lacked the time, resources, and work power to reverse the effects of the human-induced drying of the area. The fires that resulted did real damage, destroying the very attributes that made the Everglades ecologically important.

The Everglades was not the first or only park to experiment with prescribed burns. As early as 1950, Superintendent Eivind T. Scoyen of Sequoia National Park, a venerated NPS leader and a man of considerable vision, supported the designation of the Kaweah Basin in the upper Kern River drainage as a research area that would not be subjected to fire suppression. Even as he advocated the conceptual change in fire management, Scoyen asked to retain the authority to intervene if fire there threatened other areas of Sequoia.[32] The National Park Service accepted the principle that fire should not be instantly suppressed in some parts of the park system even before the controlled burn program at the Everglades began.

The decision at Kaweah Basin hardly represented a nationwide policy change. Regional director Owen A. Tomlinson took great pains to establish that the Kaweah Basin presented a unique situation, telling NPS director Newton B. Drury that the area was "so completely isolated, with unique values that depend completely on its being left alone, that such a special designation would establish no precedent." Kaweah Basin is more than 11,000 feet in elevation. Its lightning fires cannot be easily seen from lookout posts, but with the advent of aerial fire reconnaissance, fires were spotted inside the basin more easily. When Lowell Sumner, a veteran NPS biologist who assessed parks throughout the West, urged that Sequoia managers allow the basin to remain in "a natural state, free from any human interference," he argued for a piece of wild nature that had little implication for other forms of park management.[33] By succeeding with an argument about the unique attributes of the basin, Sumner, Scoyen, and Tomlinson achieved a small objective: they protected the prerogative of a research area. They did not intend nor did they make a claim for a larger use of fire in national parks.

Other instances of variance with policy with official sanction followed during the 1950s. At Pipestone National Monument in Minnesota, Superintendent Lyle K. Linch experimented with controlled burning of grasslands

Jim Agee

Jan van Wagtendonk and NPS Fire Ecologist Caroline Nobel on wilderness burn in Yosemite National Park

Jan van Wagtendonk and Harold Biswell during the first prescribed burn in Yosemite National Park

Harold Biswell. Photo courtesy of Jan van Wagtendonk.

Jan van Wagtendonk

Ground fire in lodgepole forest from bridge at Grant Village. Yellowstone National Park. Photo by Kathy Peterson, 1988.

Trees torching at Grant Village Junction—with sign. Yellowstone National Park. Photo by Jeff Henry, 1988.

Ground fire at Norris Geyser Basin. Yellowstone National Park. Photo by Jeff Henry, 1988.

Moccasin fire in the afternoon on the left and helicopter delivering water into the smoke and fire zones. Mesa Verde National Park. Photo by National Park Service.

Nighttime ground fire at Madison River with large red glow.
Yellowstone National Park. Photo by Deanna Marie Dulen, 1988.

La Mesa fire, photo taken from Far View. Mesa Verde National
Park, July 29, 2002. Photo by National Park Service.

La Mesa fire, road not identified. Mesa Verde National Park, July 29, 2002. Photo by National Park Service.

Whiskey fire, 2003. Yosemite National Park. Photo by National Park Service.

The Robert Fire outside the border of Glacier National Park, August 13, 2003. Photo by Somer Treat, National Park Service.

The Trapper Fire seen from the Highline Trail. Glacier National Park, July 23, 2003. Photo by David Restivo, National Park Service.

Portable fire pump, 1925. Photo by National Park Service. Courtesy of Yosemite National Park.

Ranger and fire equipment. Photo by National Park Service. Courtesy of Yosemite National Park.

Commissary for 200 men set up near the scene of an August 1928 fire at Merced Grove, Yosemite National Park. Photo by National Park Service. Courtesy of Yosemite National Park.

Mack Fire Truck, with Half Dome in the background, September 1, 1931. Photo by National Park Service. Courtesy of Yosemite National Park.

State Ranger Cecil Metcalf and Supt. White conferring at Maxon's Ranch, on South Fork Fire. National Archives RG 79, Records of the National Park Service Central Classified Files, 1907–1949, National Parks: Sequoia 857.01-884, box 436, entry 7.

Supt. White on Cold Spring Trail monitoring backfire on South Fork Fire. National Archives RG 79, Records of the National Park Service Central Classified Files, 1907–1949, National Parks: Sequoia 857.01-884, box 436, entry 7.

inside the park. Recognizing that the scene around the famed Hiawatha quarry no longer resembled historic descriptions of the area because the absence of fire had created a more heavily wooded vista than had existed before, Linch sought support for burns that would recreate historic conditions. Photographic evidence from the 1920s bolstered his case; even the grasslands of that era had become heavily wooded in the ensuing thirty years. At the regional office in Omaha, Linch found archaeologists and other cultural resources professionals to be supportive. In 1950, as the summer travel season began, Linch, with the help of the Pipestone community, burned grasslands inside the park. The community found nothing unusual in this practice. Farmers and ranchers in the region had long burned their lands at the end of the summer.[34]

The Pipestone experiment was unusual, but it characterized a strategy that circumvented the restrictive NPS suppression policy. Into the 1950s, many park superintendents retained considerable autonomy, with some still able to operate more or less as free agents. Countless idiosyncratic practices were common, especially at remote or less significant parks that had strong ties to local cultures. The emergence of prescribed burnings at Pipestone and other isolated parks reflected the lack of centralized authority, a historic point of pride for the National Park Service. It also suggested that compelling the proliferating number of parks to adhere to the overall NPS line on fire was more difficult than the NPS Division of Forestry anticipated. Linch was an exception. He told his superiors about the upcoming burns, couching them in terms of a debate about the authenticity of the cultural resources setting at the park. Posed this way, the use of fire was not a challenge to the status quo. It stemmed from the kind of zealous professional rectitude that marked Linch's career. Viewed as an eccentric by his superiors, Linch created leeway for the use of fire, although his example hardly qualified as precedent. Although the regional office eventually curtailed the practice, Linch was neither sanctioned nor removed from his office because of this violation of policy. Tolerance might not describe the NPS view of such activity, but times had changed since the days when Albright, Coffman, and others weighed heavily on any advocates of fire's practical use.

By this time, scientific thinking about fire had begun to change. University-trained scientists had already revolutionized their approach. At the University of California at Berkeley, Harold Biswell arrived in 1947 to teach range management. Although he had been advised to avoid precisely this controversy, Biswell promptly committed what was at the time a heretical act: he advocated controlled burning in California. Biswell had learned the practice during a stint in the South and had the good fortune to make his suggestion just as a reversal in California state policy allowed the use of controlled fires to improve range lands. He jumped into the debate at a fortuitous moment—the first time that the use of fire as a tool to shape the landscape had been seriously discussed in almost fifty years. Biswell "was a wonderful guy, and completely unabashed in his enthusiasm for fire, and its role," remembered a former superintendent of Yellowstone, Robert Barbee, who first encountered the professor during the 1960s. "We worked well together because he was a showman. He got it done; he would have done well on Madison Avenue. We had little press conferences, and we would go out and have little seminars for people where we had the media there, and that sort of thing."[35] Although the array of forces in California allied against the practice remained powerful and prominent, Biswell's program reflected a vision for a different future for fire management.

Biswell became the pivotal figure in bringing ideas about prescribed fire to federal agencies. A native of the Midwest, he learned his craft in both the West and the South. After completing a Ph.D. in botany and forest ecology at the University of Nebraska, he was hired in 1930 to work at the USFS forest experiment station on the University of California campus at Berkeley. In the decade Biswell spent there, he studied mountain meadows and woodland-grass ranges. In 1940, he transferred to the Forest Service's Southeastern Experiment Station in North Carolina, bringing the western forester's vision of fire as the enemy. There, he witnessed the Forest Service's experiments with burning in the southern pineries in 1943, a revolutionary decision for that agency. The seven years Biswell spent in North Carolina changed his understanding of the role of fire in natural communities.[36]

When he returned to Berkeley in 1947, Biswell carried with him to the hostile intellectual climate of the American West a generation of knowledge gleaned elsewhere. The South had changed him and his views, although not for the better in the view of his Forest Service colleagues. Even after he was gently cautioned by his USFS mentor, Edward I. Kotok, who in 1947 was the chief of research for the Forest Service, to stick to range management when he reached California, Biswell soon returned to studying the impact of fire on range management. He found that fire improved range land in the woodlands of the Sierra Nevada foothills, and he developed a method of burning upslope in chaparral without the assistance of fire lines. In 1951, he began to burn ponderosa pine, the most prevalent tree in California, covering almost 4 million acres.[37]

Biswell's return to California precipitated a revolution in the way federal agencies approached fire. Federal fire response had been shaped in the voluminous fires of the northern Rockies and the Sierra Nevada. In a distant mirror of the westward motion of southerners after the Civil War, the ideology of southern fire management moved west through the person of Harold Biswell. Southern practices offered a counter, even a rebuke, to existing thinking; at Biswell's urging, the scientific community increasingly seemed willing to consider such ideas. Only the diehards, the federal agencies that depended on the Forest Service for leadership and funding, retained a full-fledged commitment to suppression.

A considerable body of scientific literature argued that fire could be a useful instrument of resources management. H. H. Chapman, a southern forester in the 1910s and 1920s, led the way; after him, others developed situation-specific research that advanced the idea of controlled burning. One influential piece, forester Harold Weaver's 1943 *Journal of Forestry* article titled "Fire as an Ecological and Silvicultural Factor in the Ponderosa-Pine Region of the Pacific Slope," attracted Biswell's attention. While working for the Bureau of Indian Affairs, Weaver implemented a controlled burning program on the Fort Apache Reservation in 1950, burning more than 50,000 acres of ponderosa pine. In the subsequent two years, wildfires were reduced by more than 90 percent on the burned acreage, a rate less than one-ninth of that on land that had not been burned.[38] Clearly, in certain circumstances, planned fire could be used to obviate wildfire.

Despite the experience of the Everglades and the growing consensus among scientists that there was a role for fire in ecological settings, the National Park Service was slow to assess the possibilities of the new strategy. Visitation, not fire, was at the forefront of NPS concern in the early 1950s, and the efforts at the Everglades were anomalies. During the postwar era, visitation grew from a low of almost 7 million in 1943 to almost 32 million in 1949 and more than 46 million in 1953. Both 1941, before the war, and 1946, after it ended, had been in the 21 million visitors range. By 1949, that number had increased by more than 50 percent. No dimension of the NPS—not visitor services, not the ranger division, and certainly not fire protection—could keep pace. Inundated and overwhelmed best described the national parks; addressing the onslaught took all the limited resources at NPS disposal. Director Conrad L. Wirth encapsulated the problem for *Reader's Digest.* The NPS could not

> provide essential services. Visitor concentration points can't be kept in sanitary condition. Comfort stations can't be kept clean and serviced. Water, sewer, and electrical systems are taxed to the utmost. Protective services to safeguard the public and preserve park values are far short of requirements. Physical facilities are deteriorating or are inadequate to meet public needs. Some of the camps are approaching rural slums. We actually get scared when we think of the bad health conditions.[39]

This problem was not confined to national parks; it permeated public land management throughout the 1950s. Campers left campfires to burn out without supervision, and drivers tossed cigarette butts out of car windows and onto dry terrain, where they smoldered in brush until all too often they started forest fires. Despite an extensive fire awareness campaign, education ran up against the fundamental recalcitrance of the public. As long as suppression remained the model, the solution to the woes of heavier use and increased carelessness could only be the application of an ever-greater quantity of resources to the problem of fire.[40]

One answer for the National Park Service came from an unexpected source. In 1956, Mission 66, a ten-year program to upgrade facilities and staffing in advance of the fiftieth anniversary of the founding of the National Park Service, provided the park system with its second comprehensive development program. In the decade-long program, Congress poured more than $1 billion into the parks, which translated into an enormous impact on every dimension of the national park system. Mission 66 financed countless visitor centers and other structures, improved employee housing opportunities at most parks, paid for road construction, built campgrounds and other visitor facilities, and helped to end the public cries to close the national parks if they could not be properly maintained. Mission 66 had as transformative an impact on the park system as had the New Deal.[41]

Although Mission 66 was not specifically designed to achieve fire suppression goals, many of its activities supported a new emphasis on this objective. Suppression had succeeded when resources were available to support it. World War II had pushed the NPS away from efforts to suppress fire everywhere in the parks, restricting its activities to easily accessible places. The postwar era compelled a fundamentally reactive posture. With important exceptions—such as the Everglades, Kaweah Basin in Sequoia National Park, and Pipestone National Monument—most parks simply responded to fires and sought to put them out as quickly as possible. A major conflagration was the most threatening crisis any park faced and the most difficult to anticipate.

Among Mission 66's primary activities, building roads and trails offered advantages for fire control as a result of the improved access they created. Roads and trails made it easier to get the firefighters' vehicles and heavy equipment to the places where fires raged. Parks such as Olympic National Park routinely used the elaborate trail systems inherited from the Forest Service as the basis of their fire response strategy. Mission 66 funds lessened the strain of upkeep, making these pathways even better tools of access for fire suppression, while they still supported other purposes.[42]

In fire management, Mission 66 first exacerbated the problems of the postwar era and then began to provide a remedy for them. In fire protection, as in so many other areas, Mission 66 functioned as a more comprehensive version of the New Deal. It provided an injection of federal resources that allowed the National Park Service to respond to the changes in visitation patterns and to the increased use of national park lands. The inclusion of fire protection in what was essentially a capital development program took a number of years. By 1960, the acting NPS director, Eivind T. Scoyen, who had considerable experience with fire as a result of his superintendencies at Glacier, Sequoia, and other parks, recognized the implications of Mission 66 on the fire management program. "The Mission 66 Program to date [has] altered the patterns of use by park visitors," he informed the regional directors. "Operating programs have changed and many other factors indicate the need to restudy the forest protection program for each park."[43]

Scoyen realized that Mission 66 efforts made the parks' fire situation worse. Capital development allowed more people to visit places that previously had been out of reach, increasing the danger of human-induced fire in many new places, at the same time as lightning fires were on the rise because of the increased fuel loads that had resulted from successful suppression. An old pattern had repeated. The fault, Scoyen believed, was that existing funding schedules for fire protection did not accurately reflect the situation in the national parks. Requests for forest and fire control reserve money far exceeded the available funding, a circumstance Scoyen believed stemmed from poor planning, but in fact was an endemic problem in the NPS. Scoyen sought an overall review of fire control as the NPS practiced it. He asked the regional directors to assess the methods of detection and the operational phase of fire control and to document estimated increases in revenues.[44]

Mission 66 provided a second instance in a thirty-year period during which the NPS had enough resources to meet every whim of a superintendent or a regional director. Cleared trails had the same influence on firefighting in the 1960s as they had in the 1930s. So did the thinning of underbrush, the clearing of danger spots, and an array of other practices in which the service engaged as a reflex. Mission 66 promoted the possibility that suppression could remain viable—if only enough resources were devoted to it on a regular basis.

Two revolutionary changes altered the direction of the National Park Service. The G.I. Bill created opportunities for many to attend college, increasing the pool of trained scientific specialists. Two reports, the A. Starker Leopold Committee's report on "Wildlife Management in the

National Parks" and the National Academy of Sciences' "A Report by the Advisory Committee to the National Park Service on Research," known as the Robbins Report after its lead author, compelled a new vision of National Park Service management. The G.I. Bill created a cadre of scientists who were interested in working for the NPS at precisely the moment that the National Park Service had the resources to hire many of them. The two reports redefined the role of science in the agency, advocating not only a research agenda but an ecological vision of what national parks should be. In a way never before tenable, the service embraced science as a guiding administrative force, its officials focusing with renewed vigor on resources management as a core mission.[45] Fire management became a significant component of that obligation.

Fire science continued to move forward, largely apart from the National Park Service. Tall Timbers, a Florida research station created in 1958 to study long-term fire ecology, became the center of some of the most exciting research on the use of fire. Developed by a wealthy landowner, Henry Ludlow Beadel and his wife Genevieve Dillon Beadel, and Edward V. Komarek, a well-known wildlife biologist who operated a wildlife experiment station, Tall Timbers set a high standard. Beginning with the publication of its first conference proceedings in 1962, Tall Timbers developed ideas about fire ecology that directly contradicted the Forest Service model. Under Komarek, the visionary scientist who pioneered the use of fire during the twenty-five years he worked in the region prior to the founding of Tall Timbers, the most innovative and even heretical ideas about fire usually could get a hearing. The open, inviting climate created healthy discussion about the role of fire in the natural world.[46]

This type of intellectual innovation was new to the NPS. The service had been a tight-knit cadre since its inception, with tremendous loyalty to the values of leadership across generations of employees. There were "greenblood" families, second- and by the 1960s, third-generation members of families who had served in the NPS. The presence of Horace M. Albright, more than thirty years past his directorship but still a vital and influential force, added to the reverence for historical practices. Despite the efforts of the first head of the NPS Wildlife Division, George M. Wright, whose life had been cut short by an automobile accident in 1936, science had rarely been a particularly strong dimension of the NPS culture.[47]

In this context, the Leopold and Robbins reports set a higher standard for science in the national park system and greatly influenced the service's natural resources management policies. Ordered by Secretary of the Interior Stewart Udall and authored by a group of noted wildlife scientists, the Leopold Report recommended maintaining or recreating the original ecology of a park as a "reasonable illusion of primitive America," a goal much at odds with the manipulated nature of many parks. Believing that natural conditions should prevail, the authors suggested removing all nonnative species from parks, putting biologists rather than interpreters in charge of managing wildlife, and emphasizing the role of fire in forest regeneration, among other management practices. The Robbins Report focused on the National Park Service's research needs. The Everglades was among the parks the committee visited, and it paid close attention to that park's experiments with fire. The report concluded that the NPS should preserve national parks primarily for the aesthetic, spiritual, scientific, and educational values they offered to the public. The service's research agenda, in the committee's view, should directly support those goals. Fire was included in the mix of components the report thought valuable to the park system.[48]

The Leopold Report specifically addressed the role of fire in national parks and argued for a change in direction. With its primary focus on wildlife and habitat, the report saw fire as a tool for shaping the park environment. Pointing to the experience of the East African grasslands, where humans used fire to shape their environment for centuries, the report argued for fire as a strategy for habitat management. Controlled fire was "the most 'natural' and much the easiest and cheapest to apply," the report averred. It acknowledged that fire suppression had created conditions that had to be modified before any systematic plan of controlled burning could take place.[49]

The two influential reports combined with an internal service transformation to create an ecological strain in NPS thinking. Among the thousands of veterans who went to college, some saw careers in government as a viable extension of their military service. A small number became biologists, wildlife specialists, and other natural resources-oriented researchers. Some of these joined the NPS, many hired with funds provided by Mission 66. Where the first two generations of the National Park Service had hated fire and treated it as an enemy, these newly trained arrivals regarded it in a much more dispassionate manner. Fire was a tool they could use, they believed, and the growing body of research, much of it stemming from Tall Timbers and from Harold Biswell and his students, supported their vision. Their very presence in the service helped to spur the move toward a different kind of strategy.[50]

Leopold and Biswell's relationship greatly contributed to both the ideas in the Leopold Report and the implementation of its goals. A. Starker Leopold, the son of Aldo Leopold, was one of the nation's premier wildlife management specialists. He and Biswell both taught at the University of California at Berkeley, and their offices were just across the street from one another. They collaborated on research, profoundly influencing one another's thinking about fire. The two lunched together and jointly taught seminars for graduate students. A group of students and faculty interested in the same topics grew around them, for at a time when the University of California was acclaimed as the best public university in the nation, the opportunity to study with a pair of such natural resources luminaries could not be matched anywhere in the country. Their labs became crucibles for a new generation of fire scientists. A number of their graduate students pursued the ramifications of fire as dissertation topics. Four—Bruce Kilgore, who matriculated in 1968, Jan van Wagtendonk (1972), James Agee (1973), and David Graber (1981)—became National Park Service scientists who influenced fire policy during the subsequent generation.[51] In the long run, Biswell's impact, in concert with that of the Leopold Report, extended well beyond high-level discussions; they created a generation of scholar/practitioners who carried his ideas forward.

Implementation exploded on the NPS landscape. In 1965, Yellowstone National Park began to experiment with controlled burning. As part of the park's Elk and Habitat Management Plan, Yellowstone undertook an aspen restoration program that relied on the controlled use of fire to achieve its goals. The program had two objectives: to ascertain if the burning of a former stand of aspen, taken over by conifers, would cause the area to revert to aspen, and to see if burning would enhance the "number and vigor of Aspen sprouts in a decadent stand" as park biologist Robert E. Howe described it. The park planned a large burn, but the summer remained too wet to sustain sufficient fire. Scaling down his expectations, Howe carefully selected a location, a five-acre patch on a north-facing slope just south of the old Tower Falls road where conifers had taken over a stand of mature aspen. Surrounded by grasslands, it provided an excellent opportunity for an experiment in controlled burning.[52]

Finding optimal conditions proved difficult. By the end of September, the chance to burn seemed to have passed. In early October, Howe tried to set up the burn, only to be thwarted by a six-inch snowfall. Undaunted, the biologist kept trying. At 1:45 p.m. on October 28, in gusting winds of up to eighteen miles per hour and with a relative humidity of 50 percent, a crew under the supervision of district ranger Bohlin began to intentionally burn timber inside Yellowstone National Park. The men sprayed the conifers with diesel oil and ignited them. The fire went up into the crowns and burned about ten trees closely clumped together. As the gusty winds died down, the fire followed, and despite efforts to reignite it, the crew recorded little success. After three hours and 200 gallons of diesel fuel, they decided that the conditions "weren't going to be favorable this year for a burn," Howe told his superiors.[53]

Despite the failure to truly burn anything of significance, the very act of attempting to burn in Yellowstone represented a major step toward the idea of controlled burning. The idea that the National Park Service would throw over the historic practices of fire suppression—even under the guise of an elk management plan—was revolutionary. The importance of even indirect implementation at the nation's most important park without a sound from the powerful advocates of suppression throughout the service spoke volumes about the imminence of a formal change in policy.

The regional office encouraged the continuation of the experiment. In December, acting regional director George Baggley visited the plot with Howe and Superintendent John S. McLaughlin and pronounced that he was pleased with the experiment. "The lack of complete success should not deter you from going ahead with the program," he told McLaughlin. Baggley knew the literature on controlled fire and suggested that the very lack of fuel load that attracted Howe to the location might have made the experiment go awry. "You have a situation at Yellowstone which does not fit the book, so far as controlled burning is concerned," he continued. "Keep it up though because I certainly think the results will be worthwhile."[54]

If not quite an attack on existing policy, the Yellowstone experiment reflected the importance of the newest studies of fire emanating from Biswell's program at the University of California. Much of that work focused on California, where sequoias had always posed an intellectual problem for fire suppression. The big trees seemed impervious to burning, and ancient as they were, they forced a different calculus. Suppression was new, but the trees were very old. Asserting that the absence of fire had a positive impact on the big trees was an exercise in futility. In 1965, Biswell began research on how managing different kinds of trees in a stand contributed to the patterns of burning that spurred better conditions for the future of the trees. He also became interested in reducing the enormous fuel loads that generations of suppression had left at the base of most of the giant sequoias. Two years later, another fire researcher, Richard J. Hartesveldt of the University of California, mapped out the role of fire in the regeneration of sequoia trees in the 1967 *Tall Timbers Fire Ecology Conference Proceedings*. Biswell also contributed, assessing the level of danger and the potential risk of fuel loads in the sequoia region. Together, the two papers clearly articulated a strategy of prescribed burning among the sequoias.[55]

This new intellectual regard for fire led to serious strategic rifts in the NPS. No matter how hard they tried, many of the more senior professionals in the service could not swallow their distaste for the idea of intentionally burning vegetation. Their careers had been built around stopping fire in any way possible. Most had lost friends to fire; all had witnessed destruction of the parks—blackened forests and burned ranches, homesteads, and lodges. Many could intellectually understand the way in which prescribed fire could prevent greater calamity, but in their hearts, they believed—and everything they had ever seen had taught them—that fire was too dangerous to be intentionally let loose. Just because the Leopold Report said the NPS should burn did not mean that generations of park professionals were simply going to go along.[56]

For a number of years, the National Park Service's fire control structure did its best to ignore the ideas developed in the Leopold and Robbins reports. Although assessing fire increasingly involved scientific analysis, fire prevention still dominated the NPS perspective. By the middle of the 1960s, the service attempted to quantify the environmental conditions that influenced annual fire levels. Comparative levels of moisture began to be a feature of the annual fire reports at individual parks, with the comparison of types, numbers, and distribution of fires increasingly linked to the patterns of rain and snowfall. Yet, throughout the mid-1960s, fire management goals remained the same, and suppression remained atop the list of NPS objectives. When Glacier National Park offered a new fire control plan in 1965, it reiterated the long-standing vision of suppression so crucial to the service's formulation of its obligations: "Every fire when discovered shall be reached and extinguished as quickly as possible," the report insisted, "whether caused by man or lightning, and whether originating in a developed section or in a wilderness area."[57]

In response to the Leopold Report, Director Conrad Wirth averred that less intense suppression merited consideration, but he did not contemplate an immediate change in policy. Lawrence Cook, Coffman's successor and a long-time devotee of suppression, manipulated information to support his beliefs. For months, Hartesveldt's study of fire ecology in Sequoia and Kings Canyon, a document that threatened the viability of suppression as a strategy, lay in a pile of papers on Cook's desk. The service leadership seemed content to maintain the status quo no matter what the Leopold Report advocated.[58]

The National Park Service faced two large-scale problems with suppression. The fragmentary constitution of the national park system raised the issue of scale repeatedly. Even at its largest parks, the NPS could not maintain a fire program sufficient to meet the demands of an exceptional year. Parks were constantly overstaffed if fire did not come, but dangerously understaffed if a major conflagration occurred. Even more, fire defied any and all strategies; it was not a bureaucratic category that submitted to administrative solutions. Fighting fire was not at all like repaving roads or maintaining clean restrooms. It worked by its own impenetrable logic.

The impetus for change came from outside the service—from its friends and supporters, energized by the environmental revolution and increasingly wary of even this most benign of federal agencies. The research supporting light burning was overwhelming.[59] Even suppression's most avid proponents acknowledged that it had been a dangerous strategy. More than thirty years of application showed that it worked best when money and labor were abundant. The New Deal and Mission 66 made suppression viable, but only temporarily. And it continually created problems. High levels of fuel loads, first noticed by Colonel John White in the 1920s, resulted from suppression, and the results of the fuel-driven blazes could be catastrophic. Some agencies could live with the consequences of such overwhelming fires, but the National Park Service could not. The iconography of the parks was so significant in the midcentury United States that large swaths of apparent destruction inside their boundaries seemed a rebuke of democracy. Even more, out-of-control fires threatened people and damaged valuable resorts, hotels, and other commercial property as well as NPS facilities. After the Leopold Report, a change in strategy became hard to resist, but suppression persisted.

Yosemite maintained the existing approach of fire suppression even as the Berkeley students engaged in the research that was revolutionizing the field. Yosemite emphasized education: it trained 1,181 children in fire safety in its junior ranger program in 1966 alone. That year, the number of human-induced fires in Yosemite grew remarkably, from thirteen in 1965 to forty-one. The decade-long average had been sixteen. Attributing the increase to a greater number of fires started by cigarette smokers in heavily trafficked areas, the park stepped up its prevention programs, added new patrols, and increased the number of fire prevention posters and "high fire hazard" signs throughout the heavily visited areas of the park. Signs along roads and trails that exclaimed "no smoking while traveling" provided additional warnings.[60]

In 1967, the northern Rockies experienced a brutal fire season that reminded the NPS that the region was dangerous fire country. Beginning in July, fire broke out throughout the northern Rockies and the Inland Empire, as the intermountain area in Idaho and Montana was called. More than 5,000 fires were reported, and more than 30 exceeded 1,000 acres in size. On July 12, 131 fires burned until they were doused by rain; on August 9, 167 started; on September 6, 97 more began; and on September 21, another 167 were recorded. Before the rains came in late September, lightning alone ignited more than 1,400 fires.

When the northern Rockies burned, Glacier National Park always was a critical concern. During the spring, fire was far from the minds of park staff and regional officials. Instead, they worried about a repeat of the dramatic floods of 1964. At 5,000-feet elevation in early May, the snow pack was 152 percent of normal. Even that late in the year, flooding seemed a more realistic threat to Glacier National Park than the prospect of fire. Precipitation up to the 3,000-foot elevation level had been about normal, and prolonged dry spells had not plagued the park. The snow melted slowly, and the rains first diminished and then ended by mid-July. Through May and June, southern slopes lost their snow mantle and began to dry. By mid-June, Glacier officials could see a crisis in the making. July, August, and September 1967 became the driest months that had occurred at West Glacier since the advent of record keeping fifty-three years earlier. The precipitation of .98 of an inch was barely one-fifth of the average.[61]

Few parks were as well prepared for fire as Glacier, reflecting its fire history as one of the most heavily burned places in the system. Coffman wrote his first fire plan for Glacier, and the park's firefighting preeminence had continued. The risk of damage to property and life was always great around Lake McDonald and the occupied areas of the park. If the NPS wanted to demonstrate the viability of suppression once again, Glacier was as good a location as any.

During the evening of August 11, 1967, dry lightning crossed the Continental Divide, continued in the early morning hours of August 12, and inaugurated Glacier National Park's fire season. Fire spotters counted more than 100 ground strikes, the first at 6:25 p.m., with the first new fire reported at 7:05 p.m. The sparks started 20 new fires, burning in total more than 12,000 acres of timber. One of the most aggressive fires, the Flathead fire, was discovered about halfway up the Apgar and Huckleberry mountains. The fuels in this area resulted from the major 1926 fire in the region. By early afternoon on August 17, the fire had spread to 650 acres. It doubled in size in the next seven hours, powerfully angling up the slope. A cold front changed the direction of the wind, and by 10:30 p.m. on August 18, the fire had jumped the North Fork of the Flathead River, reaching the Flathead National Forest. It continued to spread downhill until August 20, when the center portion of the fire burned out and created two smaller fires, one on the northwest part of a ridge and the other on the southeast. By August 22, more than 4,645 acres of timber had burned.[62]

The situation turned worse on August 23. The weather bureau issued a red flag weather alert, predicting that a weak Pacific frontal system passing through the area would bring high winds and dry lightning storms during the subsequent twenty-four hours. In response, the park tried to link fire lines dug with bulldozers with those dug by hand before the winds arrived. By 3:00 p.m., before the lines could be joined, the winds accelerated to between forty and sixty miles per hour. Firefighters were forced to retreat as the fire rapidly spread. Individual fires could be found as much as a half mile in front of the main fire, started by embers thrown ahead by the force of the wind. By the end of the day, 3,500 more acres of vegetation had burned.[63]

Fires continued for another month, a result of the dry conditions. When they finally came to an end, suppression advocates pointed to their successes. Throughout the northern Rockies, fires had been controlled, a stunning contrast with the terrible fires of 1910. Instead of the roughly 3 million acres of timber that had burned in 1910, the 1967 fires covered a total of only 90,000 acres. Fatalities had dropped from seventy-eight to three, with one resulting from a heart attack. Technology played an enormous role in this success. Aerial infrared scanners, oblivious to the smoke plumes that obscure vision, mapped fire perimeters. Fires that would have burned for days in 1910 were detected early, and control efforts began within hours. Radio, telephone, and teletype networks provided instantaneous communications, allowing for immediate knowledge of new fires and coordinated responses. A national infrastructure also contributed to the 1967 success. The region was declared a national disaster area, and the federal Office of Emergency Preparedness joined in the suppression efforts. Full closure of the national forests, a status akin to martial law, was enacted, keeping visitors away and limiting the chances of additional accidental fires. The response was impressive; the damage—with the exception of the 56,000-acre Sundance fire in northern Idaho—was minimal. Suppression, most observers agreed, worked.[64]

A tone of exhaustion and impending doom permeated the accounts. Suppression advocates no longer trumpeted their successes with a vision that they were winning the fight. Instead, the fires of 1967 were another episode in a never-ending war against an opponent that could not be vanquished. The National Park Service finally recognized hard reality: fire would not go away, nor could it ever truly be suppressed. As long as the existing policies remained, the NPS would continue to invest ever-greater resources in an irresolvable conflict. Every year, it would have to scrounge for dollars and firefighters. Each successive season, the service would recount the damages, savor any victories, lick its individual and collective wounds, and prepare for the next year.

The 1967 Glacier fires also subjected the National Park Service to severe criticism. Led by former U.S. senator Burton Wheeler (D-Mont.), a longtime nemesis of the service, inholders, those who owned land inside park boundaries, charged that the NPS botched the management of one fire. It was a simple matter to put it out, some averred, and pointed to NPS mismanagement as the source of their losses. Such recriminations were typical in the aftermath of most fires, but Wheeler's status gave this set of attacks more serious ramifications. Irate complainants ignored the other thirty-five fires in the park that summer as well as two bear maulings, but the criticism still stung. Even when the NPS provided suppression, it could still be chastised by an ungrateful public. The situation offered one more reason to break the cycle and to consider new approaches to fire.

By 1967, the general tenor of scientific thinking about fire truly had changed. Fire had come to be seen as a natural force in an age that valued the concept of nature. The combination of changes in the people who comprised the agency and changes in ideas about fire brought a new approach to the forefront. Even as the northern Rockies grappled with the 1967 fire season, the end of the era of suppression was at hand.

FIVE

Allowing Fire in the National Park System

"The presence or absence of natural fire within a given habitat is recognized as one of the ecological factors contributing to the perpetuation of plants and animals native to that habitat," attested page 17 of the 1968 Green Book, the administrative policies for NPS natural areas. They were just words on paper, but their articulation signaled a revolution in the National Park Service's fire policy. Fires would be allowed to "run their course" when they remained within preestablished boundaries and contributed to management objectives. Similar changes appeared as well in the Blue and Red books, the service policies for recreation and historic areas. After fifty-one years of NPS history and nearly a century of reliance on fire suppression in the national parks, the National Park Service embraced a new vision of the role of fire and the obligations of managers in the national parks.[1]

For almost 100 years, the goal of U.S. land management agencies had been to put out every fire in every national park, indeed every fire on public land in the United States, as quickly as possible. Not only did the new policy discourage suppression, it actively encouraged prescribed burning. A new era in management had begun.

The introduction of fire as a tool in NPS policy reflected larger changes in federal philosophies and in American society in general. The generation-long pressure for wilderness preservation that began with David Brower and the Sierra Club generated a spate of new legislation at precisely the moment that the wider public embraced the new set of values labeled "environmentalism." This remarkably concentrated period of law making gathered momentum throughout the 1950s and reached an initial peak with the passage of the Wilderness Act in 1964. It culminated in the National Environmental Policy Act of 1970 and the Endangered Species Act of 1973 and continued through the five-year reauthorization of the ESA in 1978. During this fourteen-year heyday, the rules by which federal agencies governed public resources changed dramatically, allowing for far greater public input and much more sustained compliance activity. The geometry of federal fire management changed as a result, creating new alliances for the NPS and putting the historically dominant U.S. Forest Service on the defensive.[2]

The impact on fire management was enormous. As environmentalism became part of the national language and the nation confronted the "quiet crisis," as Secretary of the Interior Stewart Udall labeled the looming questions about the quality of the physical environment in 1963, federal agencies broadened their approach to fire. The Boise Interagency Fire Center was established in 1969 to unite the Bureau of Land Management's developing fire expertise with the existing prowess of the Forest Service. The result was a higher level of cooperation that led participants in new directions. The Forest Service soon modified its suppression policy, initiating natural fire experiments in 1972 that began a change in direction, following the lead of the NPS. Soon after, the move to integrate fire activities among federal agencies gained more momentum. The National Wildfire Coordinating Group first met in 1973, tasked with bringing every aspect of fire management, including multiagency fire planning and training, under one standardized rubric. In 1974, this culminated in the doctrine of total mobility, which allowed firefighting resources, especially crews, to be used by any agency anywhere. Prior to 1974, each agency had its own standards and could only accept crews and resources that met those standards. Because of a memo of understanding between the secretaries of agriculture and interior, the agencies worked toward a common standard.[3]

With its bold and at the time idiosyncratic change in policy, the National Park Service moved to the forefront in fire management. It soon came to study the impact of fire rather than to assess ways of stopping it. Although fire had been a side issue in the Leopold Report, ostensibly devoted to the ongoing elk crisis in Yellowstone, the report's vision of natural resource management loomed larger as new federal statutes began to affect the NPS. In 1964, passage of the Wilderness Act concerned NPS leaders, who saw in the new law some rules that took away their prerogative to manage land as national parks. During the 1960s and early 1970s, NPS officials watched as wilderness advocates attacked the Forest Service over questions of sustained timber yield and wild land. They recognized their own vulnerability and sought an alternative strategy. The Leopold Report provided an alternative to imposed wilderness standards that let the NPS keep the administrative discretion that formal wilderness designation overrode. Faced with two new philosophies about which it was ambivalent, the National Park Service embraced the Leopold Report over the Wilderness Act. This led directly to the new administrative guidelines for natural areas in the National Park System in the 1968 Green Book.[4] National Park Service resistance to wilderness helped to preempt its acceptance of the USFS fire suppression model, and by melding resource management and strategic goals, anticipated the direction that most fire management soon followed.

This shift toward a new vision of fire guaranteed an important change in who managed different dimensions of fire. In the model inherited from the Forest Service, most fire staff in the NPS had been trained as foresters. After the Leopold Report, wildlife biologists—often students of its primary author, A. Starker Leopold—entered the NPS and following the broad outlines of Leopold's work, began to blend fire with wildlife policy. These science-oriented researchers helped to create a model for intra-agency research and even helped to place fire science into the matrix of federal research. They communicated as easily with peers in other agencies as had their predecessors, recognizing that the physical boundaries that divided jurisdiction had little to do with ecological reality.

The influx of wildlife biologists paralleled the gradual departure of foresters from the NPS. As fire ecology competed with forestry as the model of fire management, the position of *forester* nearly disappeared from the National Park Service and along with it, ties to colleagues in the USFS and forestry in general. Attrition accounted for a good portion of this change. As hiring patterns shifted, beginning in the 1950s, NPS foresters as a group grew older, while their recently trained successors were younger and believed in the principles of ecology as they had learned in college. Over time, the change in personnel moved fire management and ecology ever closer to what would later be called ecosystem management.

During the brief moment in National Park Service history when the influence of science was at its peak, fire research enjoyed a malleability and viability that had never before been possible. Even as Tall Timbers challenged the dominant Forest Service research vision, the NPS cultivated its own fire management program.[5] NPS fire research, built from the Everglades studies of the early 1950s, expanded to the national parks in California. The critical experiments in the Golden State started at Sequoia and Kings Canyon national parks, where researchers examined the indispensable role of fire in sequoia regeneration and the threats posed by fuel build-up to the mature "big trees."[6]

Despite the NPS's interest in such programs, the emphasis on the specific needs of individual tree species detracted from the overall implementation of fire research. The desire to protect the sequoias hardly translated into an overall argument for the general use of fire, even if policy articulated such a stance. The result was slow program development. Although the Robbins

Report argued vociferously for a separate NPS fire research arm, despite valuable research conducted inside the agency, this idea did not come to fruition quickly.

When the National Park Service ceased its efforts to suppress every fire in the national parks, it upended the rationale and logic that had governed the treatment of fire for more than a half century. The path to the 1968 policy change was long and convoluted, for it represented a substantial change not only in the way the service treated fire, but in the explanations NPS officials offered to their constituents, Congress, and the press. NPS representatives argued that the National Park Service's obligations were different from those of other federal agencies, including the Forest Service and the Bureau of Land Management. Bruce Kilgore, who had been an undergraduate student of Starker Leopold's in the early 1950s, served as editor of *National Parks* magazine and later became editor of the *Sierra Club Bulletin*, where he published the Leopold Report in its entirety. The Sierra Club "thought, as an entity, that [the Leopold Report] was one of the most outstanding contributions to the management of parks," he recalled. Kilgore believed that Leopold had "total support" from Secretary of the Interior Stewart Udall for the report, a circumstance Kilgore regarded as a determining factor in the report's implementation.[7]

For the National Park Service, the Leopold Report had been the catalyst for rethinking fire, but the real energy for this shift in policy emanated from Harold H. Biswell. His research underpinned the evolving practice of controlled burning, and students gravitated to him. One, Jan van Wagtendonk, had worked as a firefighter and a smoke jumper in Oregon and Alaska before studying fire ecology. "It seemed silly to me," he recalled almost four decades later, "to be putting fires out in the tundra."[8]

Biswell played an instrumental role in the shift from theory to the practice of introducing fire. In 1964, Biswell received permission to begin giant sequoia restoration studies on a 320-acre University of California experimental tract, Whitaker's Forest, just west of Kings Canyon National Park. The Redwood Mountain grove of the big trees extended into the forest from the national park, providing an unparalleled opportunity to study the impact on precisely the tree species and forest composition that existed within the park's boundaries.

Biswell found an environment that reflected a century of suppression. He estimated more than twenty-two tons of combustible material per acre. Beneath the largest trees, he found competition from more than 500 dead and standing small trees per acre, which Biswell determined resulted from the human suppression regime. He found an additional 900 living white firs and incense cedars, mostly between one and eleven feet in height. Wildlife was absent from the area because the dense material on the forest floor eliminated most underbrush and food plants. The fuel load was so great that 4-H Clubs had ceased to camp in the region for fear of wildfire.[9]

As Biswell's research showed, burning had a positive impact on the big trees. In a comprehensive program that began in 1964 and continued until 1975, Biswell, Richard Hartesveldt, Howard Shellhammer, Tom Harvey, and Ron Stecker studied the impacts of prescribed fire. Their work demonstrated that giant sequoias depended on fire for germination and early survival. Hartesveldt trenchantly observed that, without fire, giant sequoias would become an endangered species.[10]

Biswell's efforts guided the NPS program at Yosemite as well. Robert Barbee, who considered the Leopold Report a "manifesto," visited Biswell at Berkeley "to lay some philosophical groundwork, some scientific expertise, for my resource management plan." Biswell agreed to see him and Barbee "figured I would get maybe fifteen or twenty minutes. So I got down there, and he was in his office, and he said 'well, let's go,'" Barbee recalled. "And I said 'where we going to go?' He said 'we can't sit here and talk about fire, we are going up to the experimental forest'" at Yosemite. The men spent the next two days at Yosemite exploring the prospects for introducing fire.[11]

Over a period of a little less than a decade, Biswell's fire experiments transformed Whitaker's Forest. By 1973, Biswell observed that burning had created a condition that more resembled historic time: Whitaker's Forest was more open and parklike, with fewer small trees. Shrubs and wildflowers returned, and "the forest is becoming scenic," Biswell told a meeting of the American Society of Agronomy in 1973. Biswell saw this as validation of his vision. Even more telling, he held demonstration burns in August of that year, when much of the rest of the state endured withering wildfires or lived in constant fear of the next outbreak. His prescribed fires did not burn out of control, further substantiation of the viability of a pattern of controlled burning that limited the amount of available fuel during the most dangerous fire seasons. The use of fire, Biswell believed he had conclusively shown, allowed for better management of timber, a more "natural" environment, and significantly less intense wildfire if one erupted.[12]

Whitaker's Forest was just outside the boundary of Kings Canyon National Park, but crossing that arbitrary line on a map meant more than simply traversing physical space. The national parks, the nation's sacred spaces, remained inviolable from more than the encroachment of most human development. Fire remained an equally large taboo in the parks, and securing permission to burn within them was an elaborate and drawn-out process. After a National Parks Advisory Board visit to Kings Canyon in conjunction with the preparation of the National Academy of Sciences report, many of the obstacles against burning in the park diminished. Hartesveldt received a grant from the National Park Service for a five-year research study of controlled burning. He planned to begin with controlled burns on a number of two-acre plots in August 1964, but the initial burns were delayed until 1965 as he waited for ideal weather and climate conditions. Still, Hartesveldt averred, "careful use of fire and cutting constitute a much more realistic approach than does a policy of 'hands off.'"[13]

The addition of Bruce Kilgore as a research biologist at Sequoia and Kings Canyon national parks in March 1968 accelerated the burning program. Although he was duty stationed at Sequoia, Kilgore reported to Starker Leopold, who had taken a special appointment in natural resource management from NPS director George B. Hartzog, Jr. When Leopold hired Kilgore, the professor was "semi-kind of my boss. He was more of my philosophical boss," recalled Kilgore. At Sequoia, Kilgore met another prominent fire researcher, Harry Schimke, who worked at the Stanislaus National Forest and was affiliated with the Pacific Southwest Range and Experiment Station at Berkeley. An expert in fire behavior, Schimke served as a combination conscience and advisor for those who contemplated changing the way the National Park Service addressed fire.[14]

The initial experiment in Sequoia and Kings Canyon began a revolution in the national parks. While most of the National Park Service's prescribed burning acreage was in Everglades National Park, the California parks engendered more controversy. In California and elsewhere west of the Rocky Mountains, allowing fire to burn was big news to anyone who lived near a park. Sequoia and Kings Canyon national parks instituted a policy that allowed lightning fires to burn in areas designated as appropriate for wildfire. When the first lightning-ignited fire allowed to burn in any national park occurred on Kennedy Ridge in 1967, Schimke was among those who watched. His sense of the fire's potential reach and the park's ability to control it made this difficult experience easier for park staff. The term "let burn" was assigned to such fires, but was soon changed. The idea of letting fires burn seemed too casual for public consumption, and such fires were instead designated with the more palatable term "prescribed natural fire." Even that term was controversial. In May 1971, Superintendent John McLaughlin of Sequoia and Kings Canyon, described by Robert Barbee as "a thinker and really quite an admirable character," instructed his staff to avoid the term "prescribed burn." Instead, he advocated the phrase "putting the role of fire back in the environment." Typically, the selected burn areas were far from homes, roads, and visitor services. Natural fires in them were reported and watched, usually by airplane, but allowed to burn as long as they did not exceed the boundaries of the designated area or threaten facilities. Some areas were so remote that one observer noted that if a fire got away, "no one would ever notice."[15]

Intentionally set fires were far more controversial for the National Park Service. Despite much confusion over the meaning of terms—with loose definitions prevailing, prescribed burning described activities as different as burning sawgrass in the Everglades and burning beneath sequoias in California—prescribed burning grew in importance. It also was terribly difficult to implement. Heavily traveled or developed areas challenged the very idea of letting fire run its course. The risks were enormous: lives and property were at stake even when lightning fires were allowed to burn. Before fire could be introduced as a tool in sensitive areas such as the giant sequoia groves or in similar places, considerable education and preparation had to take place. In an early experiment in 1969, along the western boundary of the Redwood Mountain grove of giant sequoias, hand-built fire lines were used to reduce fuel in anticipation of allowing natural fires to burn in a particular zone. Workers felled dead standing trees and cleared away underbrush and saplings around some of the big trees. Powerful hoses were kept near the fire lines in case fire broke outside those designated areas.[16] It seemed to some that securing conditions under which a controlled burn could take place required more effort than the burn itself.

Creating a reduced fuel zone demanded considerable effort. Typically, NPS crews burned a narrow strip to create a fuel break, usually near the top of a ridge. Workers then would drop down the side of the ridge about fifteen feet and use drip torches to burn upslope, and the fire would go out when it reached the burned area above. The process would be repeated another fifteen feet down the slope. This was tedious, time-consuming, and difficult work, and everyone who engaged in it thought of ways to make it easier; few succeeded. In one effort to speed up the process on an August 1969 burn, fire control officer John Bowdler allowed a crew to burn a larger section as a single strip instead of dividing it up into smaller sections. The fire started slowly, but gained speed and intensity as it charged uphill. Beneath some giant sequoias, the flames jumped as high as three times their previous height, with the heat damaging some of the trees' lowest branches. The experience of what came to be known as the "Bowdler burn" reminded everyone of the difficulty of managing fire even in supposedly controlled circumstances.[17]

The creation of the new policy neither included resources nor supported a system of implementation, nor did it clearly describe parameters. The policy was an objective, the articulation of a larger ideal with little practical instruction for its execution. Little service-wide coordination of the new policy ensued, for the combination of resources, leadership, and acceptance of the values it embodied were not found in the NPS in 1968. "It was just one of those things that developed topsy-turvy," Biswell student and accomplished scientist James Agee recalled.[18] Park units proceeded with fire planning in an individualized manner, simultaneously an advantage and a drawback in instituting the new policy across the NPS.

Superintendents and their staffs could design fire programs that fit the specific needs of their parks without much intervention from the national level, or they could choose to ignore the new policy. The variety of types of units in the national park system further complicated any standardization of the response. While recreational areas were governed by a policy that reflected the standards of the natural areas, historical areas functioned under more complex guidelines, which emphasized extinguishing fires that threatened cultural resources but encouraged prescribed burning. As a result, the revolution in fire management left cultural resources entirely out of the picture. No provision for addressing fires that affected historical or archaeological resources had been developed. It was as if National Park Service leaders believed that wildfires only occurred in natural areas.[19]

Critics and internal advocates accepted prescribed natural fires, but not their intentionally ignited equivalent, because they often regarded fire in absolute terms: natural fires were good; human-caused fires were bad. This set of suppositions created problems, especially in the management of cultural landscapes and in natural areas that were in fact sculpted by fire. The era's proliferating fire conferences brought this question to the fore, and it was debated endlessly.[20]

The idea of using fire challenged not only existing policy, but also the core values of many seasoned National Park Service officials and staff. The use of fire as a tool represented a change not only in what they did, but in the value system that underpinned the service. Such people "had a difficult time," research forester Jan van Wagtendonk noted in 2002. "Their entire career and belief system [had been] based on putting fires out and a bunch of hippie Ph.D.s from Berkeley come along and say 'you got to let it burn.' It [was] hard for them to grapple with that idea." The absence of a comprehensive structure to support the new policy made this transition even more difficult. Not only was the policy hard to fathom and respect, no clear guidelines to make it palatable accompanied it.[21]

The NPS's existing fire organization still derived from its New Deal roots. The Division of Forestry, headed by forester Lawrence F. Cook, Coffman's successor, handled all fire operations, and it shared in the culture of suppression and the accompanying ideas of fire protection and control. "The rumor was that Larry Cook said 'over my dead body' when he heard about any natural burning or prescribed burning," Kilgore recalled. An impasse resulted. The pressure for introducing fire had come from both inside and outside the National Park Service, but the people who administered fire in the NPS remained staunch suppression advocates. Their opponents mainly were the wildlife biologists who had become prominent in resource management. A chaotic situation in which different entities within the NPS advocated different responses proved to be a dramatic shortcoming as officials sought to implement new ideas about fire. "There was a lot of friction," Kilgore remembered.[22]

The devolution of responsibility to the park level meant tremendous variation in implementation and unusual categories of reporting. Conventional fire management continued, implemented in the same way as it had always been: education, prevention, and response. Presuppression activities continued, and parks reported fires in the same classes of size—A, B, and C—that had been standard for a generation. The most notable addition to the many annual fire reports was a new category, "Use of Fire in Resources Management," that reflected the new Green Book policy. "Fire was not used as a resource management tool this year" became a standard line in the annual fire reports for many parks.[23] It allowed the NPS to continue to function as it always had, as a fire suppression agency, while at the same time paying lip service to the new prescribed burn policy. The division between resource management and fire control exemplified the ample distance between the two philosophies. The result was a complicated management structure that invested much of its resources in conventional modes of fire control while its intellectual energy, in the form of new researchers, was relegated to what seemed an exotic form of management. Fire as a tool for resource management remained something entirely different and much less significant than overall fire control.

Even after the change in the Green Book, the approach to fire at Yellowstone in 1968 remained largely unchanged. The park reported only sixteen fires, four of which resulted from human action. The park approached these fires in the same manner as it always had; prevention, education, training, and suppression of actual fire remained the primary objectives, with an emphasis on technological support for suppression.[24] Smoke jumpers were seen as a viable response, with the West Yellowstone Interagency Air Operations Center activated on July 2, 1968. Smoke jumpers made a total of fourteen jumps on five individual fires in the park in 1968. It was as if suppression adherents believed that new technology would allow them to overcome a policy they found noxious. Older resources such as fire lookouts continued to be instrumental, with lookouts serving as the source of initial information for eleven of the park's sixteen fires. Despite the small number of fires in 1968, Yellowstone's numbers diverged from reporting trends at

Yosemite National Park. At Yosemite, human-caused fires, both intentional and accidental, slightly outnumbered natural fires in 1968. Visitor reporting accounted for most of the sightings of fire, with lookouts and aircraft playing a lesser role. But at Yosemite, the model remained as it had before the Green Book offered new guidelines, and little in its annual fire reports reflected the policy change.[25]

At the regional and national levels, the new policy earned closer attention. The admonition to implement an ecologically based policy came from above the regional offices and was used to remind and even chastise superintendents. "While the Service continues to be vitally interested in the reduction of man-made fires, the value of natural fire as an ecological factor must be recognized," observed Merle E. Stitt, assistant director for operations in the Western Region in 1968 and an advocate of the use of fire. In the context of the annual fire report for the National Park Service, his statement represented a major shift in emphasis. Stitt modified the vision of suppression, limiting it to human-caused fires, explaining that the National Park Service policy encouraged natural fires for their reinvigorating effect on the natural environment. Steve Pyne, the noted fire authority who as a young firefighter heard Stitt offer this prescription at the Grand Canyon, observed that the change was confined to words. "Almost nothing happened on the ground," he recalled more than thirty years later. Whether this statement accurately reflected the actions at Grand Canyon, it did suggest that some of those intimately involved with fire felt the need for faster implementation of the revised policy.[26]

Resource management plans soon reflected the new ethos. As statements of goals and objectives, they offered an ideal terrain to define new approaches to problems. The role of fire as a tool quickly cropped up in some of the most prominent parks. At Yellowstone, the 1970 resource management plan halfheartedly embraced the new ideals. The authors recognized that the existing principles by which the NPS managed fire were "developed in areas where timber is managed as a crop," a clear articulation of the difference between the National Park Service and the Forest Service. Yet fire posed a conundrum. "Since we lack the fire control skills to stop fire at will," the plan continued, "we can not fully adopt a program of allowable natural fires." Some fires could be allowed to burn, the report proclaimed; others would have to be addressed in the conventional manner.[27]

This bifurcation helped to ease acceptance of the new policy. Fire was so deeply ingrained as an enemy that it was nearly impossible to expect grizzled NPS veterans to accept it as a tool, even in a limited way. The division worked equally well in that it separated people by their affinity. The wildlife biologists, who had begun to take precedence, worked in close concert with one another. At the same time, the National Park Service's aging forestry cohort was as close-knit a group of professionals as any in the service. Innovation could occur—within limits—even as traditional policies continued to be implemented.

The parks that took the lead in implementing the new policy typically had been influenced by Harold Biswell and his students: Sequoia, Kings Canyon, Yosemite, and other California parks. Sequoia had been central to fire management throughout most of the twentieth century. In 1955, the McGee fire just outside and around Grant Grove burned across 18,000 acres in the foothills. In 1960 and 1961, severe fires in the park and its vicinity spurred interest in new strategies. The Tunnel Rock fire near park headquarters in June and July 1960 raged over 4,960 acres, and its suppression cost $884,931. In 1961, the Harlow fire, in the foothills of the Sierra Nevada just beyond the Yosemite park boundaries, cost more than $1 million and burned more than 43,000 acres of vegetation. In one two-hour period, the fire burned across 20,000 acres, caused two fatalities, and destroyed 105 structures. "This caused us to take another look," Peter H. Schuft, who served as chief ranger at Sequoia and Kings Canyon, observed. "Maybe we should be spending more on planning ways to stop [fires] from starting or putting in fuel breaks [prepared during the off-season] at which we could stop wild fires from spreading." The expansion of fuel breaks, places in which fuels had been removed to protect nearby property, in Sequoia and Kings Canyon followed. In 1965 and 1966, the parks received the funds to build 100-foot-wide fuel breaks, cleared of debris and other matter to a height of nine feet around Grant Grove and at the approach to Cedar Grove in Fresno County.[28]

The service's move toward a wider acceptance of prescribed burning gathered momentum. "We needed to know how to reduce the fuel [and] fire under the sequoias," Schuft remembered, "and how much actual manipulation was needed to burn safely." According to his contemporaries, Schuft was well known for his aggressive campaign to promote fire. He was purported to relish throwing lit matches out his car window along the park's primary roadways in an effort to ignite "natural fires." Schuft was a "wild man," Barbee recalled. Stories of Schuft tossing hot fuses out his vehicle window and rolling burning tires down off the roadway that encircled an area being burned abounded. "He used to drive [Superintendent John] McLaughlin crazy," Barbee said of the relationship between the chief ranger and his newly arrived superior, "because Pete looked at [the park] as a patient that needed artificial respiration. Anything goes, let's just get some fires going here. He was not very scientific."

Harry Schimke had developed a more controlled prescription for burning, and the park sought a strategy for utilizing it. Starker Leopold played an instrumental role. In October 1967, McLaughlin, recently arrived from Yellowstone, traveled to Berkeley for a meeting that Leopold held with the Forest Service at the Forest Service Experiment Station. "We were trying to brainstorm how we would begin this prescribed burn program outlined in the Leopold Report," Kilgore remembered. "There was a lot of skepticism shown, a lot of questions." Leopold took charge, telling the assembled foresters: "Hey, we came to this meeting to get ideas on where to go. We are not asking your opinion on whether we should go. We want to know what the best program is." Leopold's demeanor became bolder. "In fact, he said we are going to prescribe burn," Kilgore remembered. "The meeting turned around a fair bit."[29]

The results were dramatic. With Leopold's support and Biswell's experience, an attempt at intentionally starting a fire in pursuit of resource management goals seemed viable. By then, McLaughlin had been at Sequoia for almost a full year and knew the political terrain. He determined that the program should go forward. In 1968, at Kings Canyon National Park, an 800-acre plot on Rattlesnake Creek in the middle fork of the Kings River drainage was selected for a prescribed burn, the first authorized under the new policy. The location had been carefully chosen to minimize the chances of the fire getting out of control. Surrounded on three sides by rocks, the area was "basically a fir stand" Schuft recalled. "Schimke gave us a prescription to burn within, and we actually burned out the basin in 300-foot-wide burns starting at the top and burning down."[30]

The project was deemed a success, and the park planned more burning. In 1969, Sequoia and Kings Canyon national parks designated an area of 129,331 acres, almost 15 percent of the two parks, as let-burn areas, where fires would be allowed to run their course unless they affected park facilities. Most of these lands were above 8,000 feet in elevation; some were above the timberline. In addition, the parks intentionally burned 6,186 acres. Planners soon raised the elevation at which lightning fires were allowed to burn to 9,000 feet, engaged in prescribed burns in high country meadows, and initiated a 100-acre prescribed burn at Redwood Mountain Grove in Kings Canyon National Park, which took place between August and October 1969. To initiate each burn, fire specialists picked a mile-long, 300-foot-wide area atop a ridge and began to burn it in 1,000-foot sections. Crews first trimmed the fir and cedar foliage below a height of nine feet, dropping anything beneath that to the ground with all the snags. Then, a 2-foot-wide firebreak was built, dug down to "mineral soil," and fire hoses were located around each section. "We waited for each section to burn itself out before we started the [next] one," Schuft recalled. Crews progressed from the top of the ridge to the bottom.[31]

At the North Fork, Crystal Cave, and Marble Fork areas, the park undertook a larger and less closely monitored project. Beginning on October 29 and continuing until the snows came in mid-December, Kings Canyon crews initiated a 6,140-acre burn that occurred "without intensive preparation," assistant superintendent Jerry House noted. Pesticides, including 2,4-T, were used to kill brush, and existing roads and trails served as firebreaks. When the fires came close to crossing park boundaries, bulldozers created new firebreaks that prevented its spread. The burn reduced fuels around some of the park's most important places and also cleared a number of existing firebreaks of combustible material.[32]

House assessed the information gained in the burns, weighing the benefits against the concerns such a program raised. Questions clearly remained. Kings Canyon lacked appropriate prescriptions and sufficient fire weather data. "Much time was wasted trying to burn without results because of a lack of fire weather data," he observed. The park also needed to refine its prescriptions, aiming the type and heat of the prescribed fire directly at the desired result. Even more troublesome, House noted, was the expense. Although he was able to allot $33,984 on the prescribed burning program from the park's general operating fund, he believed that "to do the job right and get on a scheduled basis, we would need $100,000 a year for 10 years." Such a sum would give Kings Canyon a ten-person crew and a foreperson throughout the year dedicated to the prescribed burning program.[33]

Another concern vexed House and other administrators who allowed some fires inside their boundaries to burn. If lightning fires that went loose posed one kind of public relations problem, nothing was more damaging to the idea of controlled burning than an intentionally set fire that exceeded its boundaries. Control remained an enormous issue; the public could barely understand letting wildfire alone, much less starting an intentional fire that eluded control. If such an endeavor went awry, the consequences could be enormous. House recognized that supporting a program of prescribed fire was every bit as expensive as suppression and possibly damaging not only to the park's facilities but to its reputation.

By the early 1970s, the programs at Sequoia and Kings Canyon had begun to show measurable results. Between 1968 and 1971, fifty-three fires had been allowed to burn themselves out under the let-burn program before the NPS changed the name; only one such fire, in 1970, had to be extinguished. Most of the fires took place inside designated let-burn zones; nine outside those boundaries were allowed to burn themselves out. McLaughlin felt confident that this experience showed that "natural fires under conditions pertaining in the southern Sierra [Nevada] burn out after spreading over a relatively small area." He felt certain that the experience at Sequoia and Kings Canyon showed that "resource managers could restore fire to its natural role in parks and wilderness . . . in a way that is acceptable to the public."[34]

McLaughlin's support for the fire program was a crucial dimension in its continuation. He used the power of his office to support the program even when it might have been politically more viable to back away. In one instance, "one of the fires got away and burned up some stuff. There was a big review down in Sequoia which [Lyle] McDowell came out for," Barbee recounted. "It was kind of billed as a public hanging, and I went." Yosemite superintendent Larry Hadley told Barbee: "I don't want any of these goddamn fiascos taking place up here in Yosemite." Hadley "knew where I was headed, and he didn't want his own fiasco," Barbee remembered. "He said, 'you get down there and see what you can learn from it.' So, I did." Barbee was impressed with McLaughlin's commitment to the burn program and his dexterity in handling criticism of the practice. "McLaughlin defused it immediately," Barbee observed with admiration. "He stood up and said that this is such an important program that we can't jeopardize it, and he took full responsibility for what [had] occurred. That just stopped [the critics] in their tracks."[35]

McLaughlin retained concerns about the implementation of fire regimes in national parks. Although he was an advocate of allowing natural fires to burn, he was not certain of the viability of prescribed fire. He recognized the value of the program, but feared that the financial resources upon which implementation depended would be impossible to secure. This, he believed, was a dangerous scenario. New air quality standards that stemmed from the Clean Air Act of 1970 also seemed to McLaughlin to impede the use of fire. Such rules were "being interpreted to imply that the environment can not stand any more smoke of any sort or that all smoke is bad," he suggested. Despite the new law, McLaughlin did not accept the premise that all smoke was unnatural. "Smoke from natural fires has been in our environment since time immemorial, and it may well be an essential part of it."[36]

By 1971, McLaughlin had become comfortable not only with the park's new burn policy, but also with the terms under which he as a superintendent could implement it. Late in the summer season in 1971, he issued a management directive that underpinned Kings Canyon's strategy for addressing high-elevation natural fire. Using the language of the Green Book, he linked his park's resource management goals with the natural fire directive, establishing a rationale for the park's program. The plan took shape around the goal of "perpetuating animal and plant habitats in the management unit," leading to the guideline of letting fires burn, with careful monitoring, when they were above 9,000 feet in elevation. McLaughlin reserved the right of his wildfire committee and the park fire chief or acting fire chief to order suppression of a fire that threatened to spread beyond designated boundaries. He carefully delineated the areas covered in the program, producing the clearest articulation to that point of the superintendent's power in implementing the use of fire.[37]

McLaughlin embraced fire as a tool at his parks and implemented programs that helped his resource managers. Despite concerns about the efficacy of prescribed burning, which still seemed diffuse and without concrete objectives, the program flourished. Where it was implemented, it achieved a primary goal. It reduced the dangerously high fuel loads that led to intense fires that damaged sequoias. Other parks experimenting with fire programs often looked to Sequoia and Kings Canyon for examples. They were reassured by researchers who worked closely with the parks. Richard Hartesveldt of San Jose State College remained the leader at Sequoia National Park. He suggested burning around the bases of the trees prior to a prescribed burn, removing the fuel sources that caused so much damage to the big trees. Harold Biswell and his students—Bruce Kilgore, Jan van Wagtendonk, and James Agee—also helped to persuade park superintendents of the efficacy of their strategies. As the scientists played a larger role with each passing season and every positive result, the parks' shift toward the use of fire began to solidify.[38]

Despite its proximity to the research at Sequoia and Kings Canyon and its historic position as an important fire park, Yosemite had been a backwater of fire research. "When the park service has had controversy" associated with fire, Jan van Wagtendonk observed, "Yosemite has usually escaped." The transformation from backwater to forefront happened quickly. Yosemite "had some really hot shot natural resource management people," Agee remembered. By the mid-1960s, Yosemite's leadership showed interest in using fire as a tool. In January 1966, Superintendent John C. Preston invited the charismatic Harold Biswell to speak on controlled burning, bringing in the man still widely seen as a heretic to describe his work to the park's staff. Within a few years, Biswell had become a fixture at Yosemite, and his students conducted a variety of research there. By the early 1970s, he began teaching an extension course on forest fire ecology at the park. His presence encouraged further experimentation. In time, Agee believed, the Yosemite program "became on [a] par with that of Sequoia."[39]

As a result of Biswell's work and his overarching presence, controlled burning in Yosemite proceeded in a systematic and organized fashion. The scientific basis for such activity was clear, and despite strong resistance from the suppression advocates inside the park, experiments in controlled burning proliferated. In 1968, Robert Barbee returned to Yosemite after graduate training at the "ranger factory" at Colorado State University to serve as "the resource management guy there," he remembered. "I was brand new, with an entrenched cadre of people who were parks foresters and the whole traditional scene." Many were not receptive to Barbee's program. "They did not embrace me with open arms," he smiled from a distance of thirty years. "I didn't even have an office at first. Brian Harry, the chief naturalist, let me have a little desk in the library over in the museum. He was the most sympathetic guy of all for what I was trying to do." Barbee was assigned to write Yosemite's first natural resource management plan. "The chief ranger I worked for said: 'your job is to develop this plan, but don't think you are going to do anything,'" he recalled. "I am not wired that way. And so anyway I sat there for six months, did a lot of research and started to write the plan."[40]

Writing a report and implementing it were very different objectives, and Barbee had to negotiate the distance between them. "I soon realized what we had to do there was not a technical problem," he recalled. "The technical part was easy. I [had] to be an advocate, but I [had] to do it carefully so I [did not] get identified as some sort of a nut." He faced considerable resistance and moved cautiously. "I was aware that any new administrator or new breakthrough, [any] new program manager can become a casualty quickly," he mused. "So I had to kind of employ whatever charisma I had to deal with it. That was the problem." He soon was assigned an office in the headquarters building, "in the attic," he remembered. "Then I got an office downstairs with a window, and then they reorganized the park and I ended up with what used to be the foresters and the big guys on my staff. They were Indians and they were big on this whole notion of prescribed burning." Barbee's program gained momentum.[41]

He prepared to reintroduce surface fire at a variety of locations in Yosemite, including the ponderosa pine and bear clover fuels near the Wawona Hotel, the Eleven Mile road region in the park's northwest quadrant, Yosemite Valley, and the Foresta Village. At the Mariposa Grove and the Tuolumne Grove of giant sequoias and sugar pines, "there was no way you could set a fire," Barbee recalled.

> The white fir was over your head, and you would have a holocaust there if you did not watch out. So what we did in the sequoia groves is try and stimulate the fire by cutting, burning, piling the white fir and cedar. The sequoia groves [were] just too valuable a resource and we could not take a

> chance. The vegetative shift had taken place so long ago that there were no sequoias, no seedlings or anything, it was just solid white fir.[42]

Despite such a bold plan, without adequate resources, the chances of its comprehensive implementation were slim. The introduction of prescribed fire began in fits and starts. The first prescribed burns took place in the fall of 1969 at Foresta Village, and by the summer of 1970, a rudimentary prescribed burning program was in place. Biswell, Harry Schimke, and Barbee all were instrumental in its design and implementation. Harold Weaver of the Bureau of Indian Affairs, another of the progenitors of prescribed burning, visited Yosemite to assess the work. Biswell and James Agee, working on his Ph.D. with Biswell, joined him on a trip to Mariposa Grove. Weaver was impressed with the park's accomplishments. "Improvement by clearing, piling, and burning of the dense white fir understory and the accumulated flammable debris of the forest floor is an excellent project," he informed Yosemite superintendent Lawrence Hadley.[43]

Weaver continued his tour, visiting prescribed burns at the El Capitan picnic area, El Capitan Meadow, El Capitan vista area, Wawona, Foresta, and the Tuolumne Grove of sequoias. Weaver saw tremendous value in the project, for in each locale it reduced the understory and the attendant fire hazard; brought back long-suppressed flowering plants, ferns, and grass-like plants; and eliminated accumulated debris and weak trees. "I like the work at Yosemite very much," he concluded. "It has been skillfully accomplished by men who understand fire ecology and fire behavior and who know how to use fire."[44] No higher compliment could come from a leader of the discipline.

Bolstered by such praise, the move to use fire as a tool became more aggressive. A new environmental restoration plan authored by Barbee and approved in 1970 brought Biswell's ideas to Yosemite in a formal way. Prior to human arrival, the plan stated, "surface fires were one of the most important natural agents controlling the distribution of trees and meadow vegetation." Fire had been critical; many areas of Yosemite revealed natural fires about every two years over an extended period. Suppression had significantly altered the landscape, and "active management," the combination of introduced fire, natural fire, and other efforts to reduce fuel load, had become an essential step in achieving resource management goals. "Solving Yosemite's ecological problems must certainly include the use of planned fire as a management technique," Barbee believed. He selected five regions in the park for initial restoration, regarding the effort as the "initial step in re-establishing park ecosystems that have been altered by fire exclusion."[45]

The idea of an environmental restoration program melded a number of important trends in federal conservation. Barbee recognized the growing emphasis on meeting the requirements of federal statute, stemming from the passage of the National Environmental Policy Act in 1970. That Nixon era law became the basis for the entire regime of federal action on public lands, a process that the National Park Service came to call "compliance." The act created specific requirements for land managers, opening the way for the regulation of prescribed natural fire as an environmental question. At the same time, Barbee built on growing interest in the work of Biswell and his peers, bringing fire science inside the NPS loop in important ways.[46] The result was a program that almost perfectly reflected the national mood about the environment, the new emphasis on science in the National Park Service, and the growing trend in the federal government toward statutory regulation of environmental issues.

The burning already undertaken at Yosemite National Park was such a success that a significant percentage of its staff embraced the new ethos. "It is our contention that the time has come when Yosemite should break the ground for some research of our own in the ecological manipulation of resources of the park," Yosemite fire control officer Jim Olson announced in September 1970. He proposed a fifty-acre burn on the Eleven Mile area of the Wawona District, which he wanted to undertake before the fall rains arrived. Olson needed the heat that dryness would generate to accomplish his management goals. White firs were the target; only intense heat could kill the species. Olson planned control lines to surround the fire and located the burn far from roads to limit any potential consequences. "Strict adherence to prescription levels and weather forecasts will preclude uncontrolled fire out of the area," he predicted. "Research along these lines will enable us to answer some of the questions unanswered for Yosemite fuels and conditions."[47]

In plans such as the Tuolumne Meadows experimental burn in 1970, an addendum to the park's environmental restoration program of the same year, Yosemite proposed an orderly set of small controlled burns within the meadow. Barbee designated ten five-acre plots, with a scientific rationale for each. Lodgepole pine had been encroaching on Yosemite's high meadows for most of the century, and the park sought remedies to the problem. Barbee proposed a comprehensive prescription for conducting the burns, with close observation of the weather and a spot forecast from the Fresno weather station to validate the prescription and sustain it throughout the burn.[48]

After the implementation of such plans, the idea of using prescribed fire gained momentum. Fire management fit nicely into the value system of early 1970s America, an era in which many came to see the earth as an organism rather than as a canvas for human endeavor. On some levels, this new approach suggested arrogance: humans believed that they could control fire. Americans remained firm in their belief that technological solutions existed for all classes of problems. The National Park Service drank from this heady brew, but with a little more caution than those further from the frontlines.[49]

Opponents of prescribed burns remained, some prominent in the California natural resource bureaucracy. Shandon Valley rancher Ian McMillan, a member of the California State Parks Commission, adamantly opposed the use of fire, regarding the NPS's program at Yosemite as a vanguard for a similar program in the state park system. McMillan peppered California's legislative representatives with letters detailing his objections. He informed U.S. senator Alan Cranston (D-Calif.) that the burned park landscapes appeared to him as "an artificial manmade spectacle, entirely unnatural, incongruous, extremely unpleasant to view, and a flagrant violation of the concept of nature preservation on which the park was founded." When Horace Albright, eighty years old in 1970, heard of plans to allow lightning fires to burn in Yellowstone, he went straight to NPS director George B. Hartzog, Jr. "George," said Albright, "I say this with the utmost devotion to you. If you do not stop this fire policy, at least for 1972, I'll have to enter the defense of Yellowstone."[50] Albright's vehement opposition illustrated how much and how little had changed concerning prescribed fires.

By the middle of the 1970s, the National Park Service engaged in much more than an experiment with natural and prescribed fires. Most of the major national parks—including Yellowstone, Yosemite, and Grand Teton—had established programs. The NPS approved Yellowstone's first fire management plan in the spring of 1972; by 1975, it designated the entire park as a prescribed natural fire zone except for the developed areas.[51] The use of fire became an integral part of the NPS approach to preventing fire. In his day, Horace Albright could have terminated a natural or controlled burn policy with a wave of his hand. But even the reverence in which he was held and his still significant political powers could not affect fire's status in the mid-1970s.

Major conservation figures such as David Brower and Edgar Wayburn of the Sierra Club did not like the results of fire any better than did Albright. Jan van Wagtendonk took Brower into the Mariposa Grove at Yosemite to show him new seedlings that had sprouted in the wake of a prescribed burn. "I understand philosophically, what you are saying to me, but emotionally, I just can't handle the black trees," Brower told the scientist. Brower was perhaps the leading figure in American conservation—the architect of the demise of Echo Park Dam and the most outspoken proponent of nature protection of his day. Brower's sentiments were telling. For all the good fire did for ecological systems, its damage to aesthetic qualities ran hard against the vision of pristine nature that bolstered the American sense of self in the early 1970s.[52]

Still, as late as 1974, the NPS fire program had yet to encounter significant public opposition. Few fires affected visitors or nearby communities. Until such an event occurred, the NPS was operating in a petri dish—conducting an experiment that tested neither the service's commitment to the program nor its ability to address pressure from the public. The summer of 1974 changed that. A significant public relations backlash against NPS policy and procedure resulted from the management of a small fire at Grand Teton National Park. The Waterfalls Canyon fire was a slow-burning and generally visible fire that burned for more than three months before rain and snow extinguished it. It tested NPS commitment and was a barometer of the issues that could have easily derailed the use of fire in the national park system.

Before 1974, Grand Teton National Park had joined the small group of parks that were aggressively implementing a natural and prescribed burning regime. Almost from the moment of the publication of the 1968 Green Book, park officials planned for the use of fire. Implementation took longer than park leadership anticipated. The weather had been ideal for burning in 1970 and 1971, but Grand Teton managers demurred because of a lack of research data. By 1972, they decided that enough research had been completed to begin, and NPS biologist Lloyd Loope and George Gruell of the Forest Service began planning the burn. Press releases in the *Jackson Hole News* in June and August 1972 explained the program and its goals. The park prepared a fire-vegetation management plan in 1972, specifying 125,000 acres on which lightning fires would be allowed to burn. When the Midwest Regional Office evaluated the plan, officials rejected it as too dangerous, ordering the park to redesign the parameters of its program.[53]

Even before Grand Teton and NPS staff redesigned the plan, the first prescribed burns in the park took place under previously approved conditions. The 1973 program called for two prescribed burns in Grand Teton National Park. The first, a scheduled twenty-acre burn on Blacktail Butte, was delayed for the year when wet conditions slowed the drying of the understory. Crews ignited the second planned fire—a 100-acre burn on Uhl Hill—on August 28. The conditions were difficult, for the fine fuel moisture, temperature, relative humidity, and wind speed were not within the optimum ranges of the burning prescription. To compensate for the less than optimal conditions, the park instituted a series of additional protective measures. A more than one-mile-long fire line had been cleared previously; crews removed surface vegetation, and the park widened another existing fire line and extended it. A total of fifty-six people worked the fire line, including a crew with a D-7 Caterpillar earthmover standing by for rapid response. Smoke dispersal conditions approached excellent. "The forecast was favorable, manpower present, and burning commenced at 1300 [hours]," the official report of the burn recorded.[54]

Weather conditions did not cooperate with the plan. Smoke from the fire drifted east and then southeast toward the Mount Leidy Highlands, large sedimentary mountains east of Jackson and north of the Gros Ventre River. Only a small amount settled in the Buffalo Valley, but planners feared that any smoke there would spark local resentment. The fire consumed nearly thirty acres of sagebrush that surrounded a stand of aspen, but it lost its impetus in the green and moist aspen understory. Only a few small areas burned at all, and most did not burn thoroughly. Staff members patrolled the fire area for three days, engaging in only a little additional work. On September 1, rain and snow extinguished the fires.[55]

Despite some negative results, the Uhl Hill fire gave the park confidence in the prescribed burning program. In the most basic of terms, Grand Teton had started a fire, monitored it, and achieved some resource management goals without incurring the ire of the community. While less acreage than intended was consumed, the fact that the burn came off without significant problems was a major achievement; it showed that intentional burning did not have to affect community-park relations. Nor did it have a negative effect on natural resources. Grand Teton staff members saw it as an important starting point for a regime to use fire to shape the natural environment.

Devising an acceptable plan to govern the overall burning process proved to be more complicated. In July 1974, Grand Teton National Park circulated a revised draft environmental assessment of its proposed fire-vegetation management plan. When the park held hearings on the proposal, it received an array of comments. The majority of the negative comments focused on the oldest of national parks issues: as seasonal research biologist Dale Taylor expressed it, the question was "when are things *natural* and when are we *gardening*?" Prominent wilderness advocate Adolph Murie agreed. "We should be guardians not gardeners," Murie had written in a critique of the Leopold Report in 1963; in 1974, he remained unconvinced. University of Wyoming zoology professor Oscar Paris supported natural fire but urged that "any program of prescribed burns in the park be dropped forthwith," predicting "blemishes on the land." Louise Murie, Adolph Murie's wife, observed that the many spot fires constituted more than the "minimum of management" that was consistent with park policy. The less numerous positive comments came from those with some background in natural resource or fire management. U.S. Forest Service retiree Richard E. Baldwin believed that the idea of a burning program was a good one, but that the NPS needed more fire suppression and fire behavior expertise on its fire management team. Such specialized commentary amounted to a refinement of NPS practices rather than opposition.[56]

Nine days later, on July 17, 1974, lightning kindled the Waterfalls Canyon fire. That summer around Grant Teton National Park had been extraordinarily dry, with successive months of below-normal precipitation. When the fire started in the area of a 5-acre burn from the previous year, the park staff's initial reaction was to let it go and monitor it. With other natural fires burning in both Grand Teton and Yellowstone national parks, the Waterfalls Canyon fire did not seem exceptional. It spread "slowly, sometimes invisibly, sometimes with billows of smoke. . . . At no time was the fire's behavior unpredictable or was the fire uncontrollable," chroniclers noted. Well-prepared park interpreters responded to queries from visitors by explaining the service's natural fire policy. The fire smoldered, only expanding to 200 acres after six weeks. On September 10, almost two months after it began, the fire had grown to 500 acres and was moving upslope. After that, it grew rapidly. On September 18, it reached 1,500 acres. The next day, the fire topped 1,900 acres. Its smoke filled the town of Jackson, Wyoming.[57]

Jackson had been uneasy before mid-September. The blaze was two months old and following it had become local sport. But no one in town could recall a fire that lasted as long. Whatever support existed for the policy of allowing natural fire to burn diminished daily. The "smoky pall," as one local newspaper called it, crystallized negative sentiment. "Has Smokey the Bear Become Smokey the Firebug?" a September 19 paid advertisement screamed at the townspeople.[58] The town did not object to natural fire per se; what residents said they disliked was what they perceived as a cavalier attitude on the part of the National Park Service. From the town's perspective, Grand Teton seemed content to let the fire burn itself out no matter how it inconvenienced locals. Two very different dimensions of the National Park Service's obligations—resource management and community relations—collided head on.

The controversy soon attained national dimensions. The Rocky Mountain regional NPS office issued a press release that explained fire policy while dexterously avoiding the words "let burn." Service officials had decided that the term connoted a lack of monitoring, when the hallmark of the NPS program was close attention as any fire burned. The September 20 press release was widely circulated in an effort to explain the NPS reaction to fire to the public. In early October, the *Denver Post* ran a major story on the

Waterfalls Canyon fire, and its *Empire* edition, which covered the northern Rockies, featured a photograph of the fire on its cover. On television, CBS and NBC showed filmed reports later in October, and *Time* magazine also covered the story.[59]

The National Park Service recognized that the traveling public had witnessed the fire in large numbers. By one count, at least 100,000 visitors saw at least smoke from the fire, and service interpreters scrambled to explain why the NPS did not act to suppress it. Most people could not fathom letting a fire burn, and many expressed concern about the aesthetic impact of the fire. Others questioned its impact on wildlife, air and water quality, and vegetation.

Waterfalls Canyon illustrated a problem with which the National Park Service had yet to grapple. While the NPS burn program had genuine empirical grounding, the American public feared fire. Raised on Smokey the Bear and Bambi, Americans typically thought of national parks as beautiful vacation lands. Most simply could not comprehend the need to allow fire.

Waterfalls Canyon taught the NPS the value of communicating its science policy to the public. Ultimately, the idea of controlled burning received a powerful boost from the Waterfalls Canyon fire. An article on the fire in *Time* magazine perfectly reflected the new NPS fire policy. Smokey the Bear was "no ecologist," the article asserted. "He is not aware that natural—as opposed to man-made—fires are good for forests." Even as it acknowledged opposition to the policy, in a careful synopsis of the work of Kilgore and the other ecologists in the service, the magazine embraced the use of fire.[60] The coverage refuted the local critique of the fire, a major triumph for advocates of the use of fire in the National Park Service.

In the aftermath of Waterfalls Canyon, the NPS built on this triumph with an extensive campaign to educate the public about its use of fire. "We've got a major problem in explaining our position to the public," Bruce Kilgore told *Time*, but buoyed by the magazine's positive coverage, the NPS formulated a public education campaign. Kilgore identified the single most complex problem: convincing people that fire could be a valuable tool for protecting, not destroying, their national parks.[61] In December 1974, the NPS took an important step in bringing this issue in front of the public. A three-page press release—more a newspaper article than a conventional public relations missive—spearheaded the campaign. "National Park Service Studies Show How Forest Fires May Help Preserve the Parks," the headline announced with certainty. The release attempted to dispel what it called the dominant "fires-are-bad" construct. Recounting the history of natural and prescribed burning from the Everglades through Sequoia, Kings Canyon, and Yosemite, the NPS document highlighted the ways that fire served as a positive force and as an ecological balance system for park land.

Kilgore's three-part fire program had become policy and had been put into effect in twelve national park areas: the Everglades, Sequoia, Kings Canyon, Yosemite, Grand Teton, Yellowstone, Rocky Mountain, Wind Cave, Carlsbad Caverns, Guadalupe Mountains, North Cascades, and Saguaro. With the exception of the Everglades, these parks were in the desert Southwest or in the mountains of the West. Guadalupe Mountains and North Cascades were both new additions to the park system; conceived as smaller latter-day versions of nineteenth-century national parks and lacking the long history of firefighting so common in the park system, they were easy candidates for implementation of the new NPS strategy. Carlsbad Caverns, Wind Cave, and Saguaro shared a different set of attributes: all had plenty of easily combustible resources and significant histories of suppression. They too offered the service good places to experiment with its new policy. Additions to the program in the near future included Glacier, Grand Canyon, Isle Royale, Redwood, Lava Beds, and Point Reyes. The service designated more than 3 million acres where natural fires were allowed to burn as long as they did not threaten human life or developed areas.[62]

Despite the fact that the burn program was more than six years old in 1974, the press release was the first example of a full-fledged announcement of the program as well as a push for its acceptance by the public. The National Park Service was committed to the new program, and it expressed its commitment in a concerted and consistent manner. For the first time, the fire management program had the direct support of the National Park Service director. Gary Everhardt, who ascended to the NPS's highest post upon the unceremonious departure of Ronald Walker in early 1975, had been superintendent of Grand Teton, the scene of the Waterfalls Canyon fire the previous summer. Everhardt had stood by the burn program despite its negative publicity, and he carried that commitment with him to the top NPS job. In a March 13, 1975, memorandum, issued less than a month after he took office, Everhardt informed the service of his "personal interest and involvement in the fire management program." When challenged by congressional representatives on the burn policy, Everhardt stood his ground.[63]

The National Park Service had been struggling for models from which to develop a servicewide structure for fire management. A national process was codified in 1976. Staff directive 76-12 set out to clarify terminology and procedures for the burn program. The document articulated the NPS's newfound primacy in fire management, saying, "The incorporation of natural ignitions from lightning into management programs has been led by this Service." The directive explained the change in terminology from "fire control" to "fire management," a semantic redefinition that "more accurately reflect[ed] the increasing complexity of the Servicewide program," and it formally linked fire management to the emerging field of resource management.[64] The National Park Service sought to blend fire suppression, management use of fire, and research into "a cohesive program to perpetuate the resources entrusted to park management." With its set of standardized definitions, the document outlined NPS formats and procedures for fire reports, articulated different funding sources that could be used to pay for fire management, mandated fire management plans for every area in the system, and presented a skeletal outline of how they should be organized. It also developed qualifications and staffing protocols, and it outlined fire research programs at the individual park level.[65]

The National Park Service had an agreement with the Forest Service to include its staff in the Interagency Fire Qualifications System, a rating system that allowed for the standardized evaluation of personnel from different agencies, but the NPS also needed its own standards specific to its mission. Based on the work of rangers Art Partin, Larry Bancroft, and others at Sequoia and Kings Canyon national parks, the proposal identified nine distinct jobs with specific duties; the attributes necessary to qualify included physical fitness, experience, and subjective traits such as good judgment and observational ability. Courses of study necessary to qualify also were outlined. Fire management had reached a first stage of maturity.

The maturation process led to the continued assessment and refinement of NPS fire management policies. A task directive signed by director Gary Everhardt on November 1, 1976, commanded an assessment of past and current practices, consideration of alternatives, and the development of a recommended course of action and a schedule for an efficient servicewide fire management program. It announced an additional $1 million in annual emergency funding to supplement the existing program, which expended more than $1.3 million annually for prevention. The directive anticipated a reformulation of the fire management program during the subsequent fire years to allow it to account for the differences among the 287 national park units. It created a task force that included many of the leaders in NPS fire management—including David Butts, John Bowdler, Robert Sellers, William Colony, Larry Bancroft, and others—who were expected to design a comprehensive operational program and to develop specific recommendations to develop the complete structure requested.[66]

Early in 1977, a new staff directive, 77-1, further elaborated the structure of fire management and the division of responsibilities associated with it. As a result of the task force created under the earlier staff directive, the

NPS took two major steps. The service created a direct line for fire management in the administrative structure, and interagency cooperation became a primary concern. Neal Guse, division chief of natural resource management in the NPS Washington office, was designated as coordinator of the program. David Butts, also of the Washington natural resources management staff, was selected to oversee team leader Robert Sellers. John Bowdler was assigned to the Boise Interagency Fire Center (BIFC), the ten-year-old interagency endeavor to centralize resources and strategies for addressing fire, to develop NPS-BIFC, coordinate NPS training, and serve as liaison to other federal agencies. These steps were a prelude to better integration of the fire management program in overall NPS management.[67] The Forest Service and the NPS worked to develop a shared nomenclature as interagency cooperation became increasingly crucial.[68]

Such efforts ultimately led to NPS-18, the National Park Service's Fire Policy Directive, which in 1977 superseded every existing policy for fire management in the National Park Service. NPS-18 represented a compendium of the ideas and strategies learned in the decade since the 1968 Green Book included the use of fire. The new document was complex, covering a vast array of contingencies, for NPS leaders recognized that a combination of management strategies and implementation procedures was necessary to create policy for almost 300 disparate units. The policy reiterated clear principles: as in 1968, NPS priorities were to protect lives, facilities, and cultural properties and to preserve natural resources and habitat. It articulated clear guidelines for prescribed burning. It also separated fires into two categories: management fires, which were allowed to burn, and wildfires, which were to be suppressed.[69] The NPS terms had changed, but the rules were consistent.

The prescriptions in NPS-18 were clear and direct. "Fire control" was removed from the NPS lexicon and replaced by "fire management." "Control is but one of the many appropriate parts of fire management," the policy intoned. It also located fire management inside the national resource management administrative structure, which represented a major shift from the individualized and idiosyncratic practices that had been implemented by superintendents in each park. NPS-18 also compelled each park to evaluate its fire situation and to create a fire management plan. No longer could a recalcitrant superintendent simply ignore fire issues. Even small historical units were required to develop such plans, albeit not always with the celerity or scope that were expected of parks with long histories of fire. Even parks with what NPS-18 described as "100 percent landscaped vegetation" were required to complete a fire plan. The only parks exempted were those with no vegetation—typically small urban memorials.[70]

NPS-18 fit the National Park Service firmly into the Department of the Interior's organizational structure for fire, for the addition of the NPS to BIFC brought the service in line with the rest of the department. The result was a powerful shift in the Interior Department's fire operations, which compelled the Forest Service to develop a parallel policy. That agency, long committed to suppression, accepted fire management and the use of fire in the 1978 *National Forest Manual.* Fire management, NPS style, became the dominant mode of federal fire response.

With NPS-18, the National Park Service divided fire into three distinct categories—human induced, natural, and prescribed. A management system was established for each, with checks and balances and objectives to be obtained. It seemed an ideal system, clearly defined and focused, which would allow the NPS and other federal agencies to maintain control of fire. Fire managers could be forgiven if they believed that they had triumphed not only over arcane policy, but over nature itself. They firmly believed that they had the science right, that they understood fire and its circumstances and had devised strategies to make it work for all the national parks. As with any policy, a test would soon come.

SIX Managing Fire

As the National Park Service began to explore its new models of fire management, it faced the peculiar realities of its situation. National parks were psychically important, but they comprised very little of the land that needed to be managed if something other than suppression were to become the norm. The national parks were a boutique setting; it was easy to experiment there, but the conclusions that resulted were not necessarily relevant to the larger questions of fire management on public land in the United States. For national park fire management policy to remain significant required coherence at the upper echelons of the agency. That stability was lacking at a crucial moment in national park history.

During the late 1970s, the National Park Service experienced significant upheaval that refocused its mission in new ways. Changes at the top that began with the appointment of Ronald Walker as NPS director in 1973 continued, with a rapid succession of directors following throughout the decade. A genial, calm, and cautious man, Gary Everhardt, who succeeded Walker early in 1975, found the director's post to be more intense than any previous management post he had experienced. He left within two years, succeeded in 1977 by William Whalen, the superintendent of one of the premier urban national parks areas, Golden Gate National Recreation Area. Whalen himself was replaced not three years later, in May 1980. No prior director, with the exception of Arthur E. Demaray, a long-time associate director who was appointed as director for nine months prior to his retirement in 1953 in a gesture of appreciation for his long service, had served fewer than eight years. In the view of much of the NPS line staff and many observers, the top post in the NPS had become politicized, a demoralizing turn of events that dramatically altered more than a half century of apolitical leadership of the National Park Service.[1] This lack of continuity at the top hurt fire management at precisely the moment it moved toward institutionalization within the service.

The turbulence that existed within the fire program stemmed from a number of structural issues. Implementation of NPS-18 lagged behind its philosophical statement as energetic and enthusiastic fire managers applied their ideas to individual parks. Many surprises resulted, from both planned fires and natural ones. Despite valiant efforts to design and implement fire policy for the entire park system, the newly designed process remained tenuous.

The idea of the national parks as "vignettes of primitive America" and the emerging effort to define designated wilderness under the Roadless Area Review and Evaluation (RARE) process made the NPS position defensible.[2] The shift was called "process preservation"—protecting the ecological processes, not the vistas that visitors craved. The widespread preoccupation with the idea of "natural" lands—as a social construct rather than as an environmental condition—easily included the use of fire. In the choice among types of fire, "prescribed natural fire"—fire ignited by lightning in zones where fire was allowed to burn—was widely seen as the most desirable, as human intervention only muddled the fires' ecological purity. Wilderness provided the template for "natural." This perception was of a piece with the times, idealistic rather than pragmatic, yet it had a powerful influence on policy.

Tensions arose between the local and national levels based on the planning process and the amount of land that was to be burned. Into the early 1980s, most intentional burning on NPS lands took place in Florida at Everglades National Park and at Big Cypress National Preserve, established on October 11, 1974. The conditions in Florida and the region's cultural history made the use of fire a common and accepted practice, so common that the state's regulatory process seemed to encourage rather than restrict burning. By the end of the 1980s, the state of Florida certified burn managers—some of them private citizens—to conduct prescribed burns; these people were not liable for damage caused by a prescribed fire as long as they followed state guidelines.[3]

Elsewhere, prescribed burning remained problematic. In many instances, prescriptions—the predefined conditions under which fire was allowed to burn—were loosely or poorly defined. Despite the service's emphasis on planning, there were no national standards or models on which superintendents could rely. As a result, prescribed burning programs proliferated with too little planning and without clearly defined parameters.

Still, across the national park system, fire management planning moved to the forefront. An astonishing number of plans were written in the late

1970s and early 1980s at parks as diverse as Mammoth Cave National Park and Antietam National Battlefield. The new emphasis on fire plans produced a higher caliber of document than had ever before been possible in the National Park Service. Leading scholars played an instrumental role; in many ways, the concepts in the plans stemmed from their research. At Pinnacles National Monument, James K. Agee and Harold H. Biswell coauthored the park's first comprehensive fire management plan. They presented the plan at the First Conference on Scientific Research in the National Parks in New Orleans, Louisiana, in November 1976. Their document assessed the evolution of fire practices at the 14,500-acre monument, showing the ecological consequences of suppression and the need for a program that used fire as a tool. Sophisticated in approach and cognizant of contingencies, Agee and Biswell's plan set the standard for the NPS.[4]

Following the lead of scholars such as Biswell, Agee, and Jan van Wagtendonk, the parks with the most difficult fire histories developed fire management plans of remarkable flexibility and versatility. Glacier National Park's document revealed the complicated evolution of its fire planning. The 1977 fire management assessment provided the most comprehensive analysis of conditions in the park's history. Embracing the idea that fire is a natural force, the park sought strategies that would "perpetuate Glacier's wilderness, with the greatest safety for residents, visitors, and non-park property."[5] Following the organizational model that developed in response to the requirements of the National Environmental Policy Act of 1970, especially the environmental impact statement process, fire management plans expressed the preferences of managers in the language of alternatives. After describing the problem and the conditions under which management took place, parks offered alternative management plans and their expected consequences.

The caliber of alternative management plans varied. While Glacier National Park produced a remarkably sophisticated and comprehensive analysis of its fire situation, many other parks lacked the combination of resources and sense of necessity to invest as heavily. The plan that resulted from the Glacier assessment, approved in June 1978, articulated conditions under which prescribed natural fire and artificial ignition would be utilized as part of the park's regime. The four alternatives were reviewed and rated. Allowing every fire to burn was clearly impossible, if for no other reason than the inordinately high frequency of fires that would result. Complete suppression was beyond the park's reach; it was also counterproductive to offering visitors a glimpse of the landscapes they might have encountered as early American pioneers. The only alternatives that provided answers to Glacier's fire problems permitted the use of both natural fire and ignited fire along with suppression. The plan targeted "certain critical sites," such as ponderosa pine stands, for prescribed fire as a component of "maintaining a sound, natural ecosystem." It permitted natural fire where "values at risk [were] minimal." Yet the plan carefully deferred to the expectations of its neighbors. "Any action other than total suppression," the report read, "requires a review and endorsement by a Fire Management Review Team." This effort to assuage public concern reflected the realities under which the National Park Service operated, which some park plans failed to address.[6]

The planning process triumphed at Glacier National Park. With the acceptance of the fire management plan, one of the most difficult fire parks in the system had a flexible administrative structure for addressing conflagrations. The plan allowed park officials a tremendous amount of leeway in decision making, with a powerful emphasis on those decisions that had become the hallmark of fire management. The new values—those of the use of fire—became the rule rather than the exception. The National Park Service had developed a fire management structure that seemed as able as suppression had once been to address the problems of fire at a major park.

Yet planning was only half the equation. Even after two University of Montana professors, James R. Haback and Robert W. Steele, received a $49,943 grant for a three-year fire ecology study in 1980, Glacier National Park was unable to undertake any proposed prescribed burns. The window of opportunity for burning was always extremely short at Glacier, and in 1980, the ground cover was only dry enough to carry fire during one week. "Unfortunately we were thwarted by a series of crises over which we had no control," acting superintendent Joe Shellenberger told Haback. The park endured a grizzly bear crisis that year that required the deployment of all available park personnel. Three fatal maulings were reported inside the park, the last just prior to the scheduled burn. Two female grizzlies were trapped inside Glacier; another was caught outside its boundaries. Nine grizzlies were counted in the West Glacier–Apgar area. "There was simply no one available to conduct a fire management burn," Shellenberger informed the disappointed professor. Even though the research was "of some urgency," the project was temporarily derailed. Two years later, supported by the research of Ron Wakimoto at the University of Montana in Missoula and of Bruce Kilgore, during a four-year stint as research project leader at the U.S. Forest Service's Northern Forest Fire Laboratory in Missoula, Glacier National Park did finally prescribe burn a ponderosa pine forest in the North Fork of the Flathead drainage in September 1983.[7]

Yosemite, which had become the service's premier fire management park as a result of the efforts of van Wagtendonk and others, also developed preeminent fire planning. By the late 1970s, van Wagtendonk's work had laid the basis for a revolution in the park's planning. His papers outlined the achievements and the consequences of fire management, showing how Yosemite had measured its prescriptions for fire and how science changed the parameters of such planning. This work was reflected in the park's 1979 fire management document, "Natural, Conditional, and Prescribed Fire Management Plan." "The forest has become increasingly susceptible to catastrophic wildfire as both living and dead fuel loads continue to increase," the plan claimed. "The absence of the open park-like forest described by early explorers in the Yosemite region has resulted in the visual impairment of the natural scene, and consequently has decreased the value of the Park experience for many visitors." This statement of the problem nicely summarized the need for scientific management to merge with the visitor experience.[8]

The idea of conditional fire management—seasonal adjustment of which areas would be allowed to burn—allowed a new dimension in fire planning. At Yosemite, the fall months allowed greater management leeway, and under clearly described conditions, units that encompassed lower and upper mixed-conifer and red fir would be allowed to burn. Conditional fire management would terminate on December 31 each year to eliminate fires that might carry over to the following spring.[9] The Yosemite plan devoted a tremendous amount of energy to clearly defining the boundaries between conditional and natural fire. The description of the boundaries contained a detailed and accurate assessment of fire boundaries that exceeded the norms previously established for prescribed burns.[10]

This strategy affirmed complete confidence in the management of fire by science. Increasingly at the center of the fire management revolution, Yosemite showed the most direct influence of Biswell and his students. Its plan was unambiguous, articulate about the science behind fire, but seemingly purposefully tone deaf to the public issues relating to fire management.

Sequoia and Kings Canyon national parks also had been at the center of the revolution in fire practice, and they produced one of the most comprehensive and sophisticated studies of the first generation of fire plans. Sequoia and Kings Canyon's complicated fire history and the long history of record keeping provided some of the best information in the NPS for evaluating the role of fire. Few other parks could produce a chart that showed fifty-five years of fire history, with the frequency of fire categorized by origin.

The decade-long history of prescribed natural fire provided another dimension that many parks lacked. By 1979, the parks had started 155 fires, which together burned across 19,730 acres. While the total paled in comparison to the Everglades, it far surpassed any other park in the system.[11]

The Sequoia–Kings Canyon plan developed the most revolutionary dimension in National Park Service fire policy: the idea of introducing fire to areas where suppression had altered the natural fuel load and the composition of vegetation. The plan considered the option of allowing some human-caused fires to burn, a perspective easily regarded as a contravention of NPS-18, which required human-induced fires to be controlled to protect parks from unnatural ecosystem change and to prevent damage to property and lives.[12] The most powerful claim in the plan for innovation stemmed from the vast base of prescribed burn research that underpinned its contentions. It offered a twelve-year controlled burning plan, designated by area and size, with the total acreage of proposed burn areas ranging from highs of 19,758 acres to lows of 4,471 acres. The park's thorough, scientifically based research was persuasive.[13]

The fire problems of parks such as Sequoia and Kings Canyon were enormous and merited the immense investment in resources they received, but their situation did not mirror that of most of the parks in the system. The NPS found itself in what was a characteristically uncomfortable situation: investing vast quantities of national funding into particular parks that were extreme examples of one problem. This approach was entirely necessary, for it would not do to allow huge fires to devour the nation's treasured parks, but it also created policy on the basis of anomalies.

One of the boldest programs was burning around a number of big trees at Sequoia–Kings Canyon, a symbolic statement of immense proportions. A resident of nearby Three Rivers, California, Eric Barnes was outraged by the char left by prescribed fire on the big trees and complained to Senator Alan Cranston (D-Calif.). Bruce Kilgore drafted the response, explaining the ecological advantages of such fire, and helped to empanel a seven-scientist committee headed by Norman Christensen of Duke University. The park suspended prescribed burning while the committee studied the program. In 1987, Christensen's committee released its report, generally supporting the controlled burning program. In the committee's assessment, the ecological value of fire superseded aesthetic considerations. The committee recommended two types of prescribed burns: restoration fires to reduce fuel load and simulated natural fires to mimic historic natural fire patterns.[14] Yet despite the affirmation of the parks' ideas, the experience taught NPS personnel an important lesson: public involvement was necessary if the service were to find public support for its policies. The service realized that fire science and strategy had to be tempered by living in the real world, and throughout the NPS, superintendents, regional directors, public relations people, and everyone else finally understood this.

But most parks could not invest resources in the level of planning engaged in by Glacier, Sequoia, and Kings Canyon. Urban parks, historic homes, and other similar parks had more to fear from building fires than from wildfire; some city parks lacked anything more than a front lawn.[15] Parks such as the Thomas Stone National Historic Site in Maryland, the Edgar Allan Poe National Historic Site in Pennsylvania, and the Tuskegee Institute National Historic Site in Alabama had neither the need nor the structure to produce fire management plans of the caliber of Glacier National Park's plan. Immediate suppression was necessary and expected at such parks; there was nothing to gain from allowing natural fires to burn, and intentional ignition bordered on prosecutable pyromania.

Some parks were surprisingly cavalier in their response to the dictate to produce a fire management plan. A prominent example was the Custer Battlefield National Monument, later designated as Little Bighorn Battlefield National Monument. Where Agee and Biswell developed a 70-page document for Pinnacles, and Glacier National Park expended more than 200 pages on a mere assessment, a prelude to a fire plan, Custer Battlefield assembled a 4-page plan that displayed minimal comprehension of the idea of fire management, the use of fire as a resource management tool, or any of the other innovations that had followed the revolution in NPS thinking about fire. In essence, the park submitted a plan that advocated suppression. It listed resources, fire equipment, the details of a training program, an explanation of its reliance on the Bureau of Land Management (BLM), and a description of the fire crews available both winter and summer.[16] Whether Superintendent Richard T. Hart simply did not track the changes in fire policy or whether Custer Battlefield's unique circumstances—its position as a cultural site and de facto cemetery—dictated a different vision, the fact remained: the fire management plan submitted in 1977 was more reminiscent of an earlier era.

Yet Custer Battlefield became a test case of the impact of fire on park cultural resources. In 1983, a fire swept up Deep Ravine and burned the park's grasslands. The 1876 battlefield had experienced fire suppression for a very long time—some accounts suggested it had not burned since the battle—and the complete burn of the historic section of the park provided the first opportunity to use modern archaeological techniques to reassess the battle's historic scene. NPS archaeologists spent the 1984 season on the newly cleared field, uncovering bullets, skeletal remains, metal fragments, and other remnants of the battle, all of which helped to redefine the historical understanding of what had transpired. Out of this work emerged an interpretation of the battle that more closely coincided with the accounts of Lakota and Cheyenne people than with the myth making of that marked the first one hundred years after the battle. A different understanding of Indians' use of weapons, the flow of the battle, and other dimensions of the story resulted.[17]

The Custer Battlefield fire and its impact on historical knowledge opened another area in which fire could be transformative. Although Mesa Verde National Park had experienced archaeological discoveries as a result of fires, the idea that fire could contribute to resource management had not been in the forefront of NPS thinking. The situation at Custer Battlefield provided a new appreciation for the role of fire in other types of NPS management situations.

In December 1981, NPS director Russell Dickenson committed the service to a new program, FIREPRO—an operations analysis and budget management process modeled on similar programs in other Department of the Interior agencies and in the Forest Service. It utilized a common process to enable land managers to systematically analyze and quantify fire management needs and to provide a baseline for appropriate levels of personnel, training, equipment, and supplies to achieve resource management goals.[18]

FIREPRO sought to protect cultural and natural resources by assessing the level of risk to each and deploying resources based on that risk. The program treated the potential for wildland fire in proportion to its historic rate of occurrence, shifting the focus away from the ongoing development of fire plans for urban parks and other places where NPS resources were not likely to play a major role in fire planning and response. FIREPRO established three levels of response, with Level III denoting the highest level of occurrence and danger of repeated fires. In essence, FIREPRO attempted to balance response through the division of resources ahead of conflagration rather than in its aftermath. This simultaneously stabilized the NPS fire budget and let parks plan for emergency situations without depleting their ordinary operations budgets.[19] FIREPRO let the NPS address the perennial lack of funding; at the same time, it redefined fire as a national phenomenon rather than a local or regional one. It was also an attempt to secure more funds for NPS fire management. The core account, called PWE 342, was designed to be used only for emergency funding, but the efforts of adept administrators created a situation in which national parks used these funds in lieu of their regular budgets.

Fires continued to thwart the best efforts of planners, for the emergencies often fell well outside of the categories that the NPS could devise to contain them. In the summer of 1977, a fire at Bandelier National Monument outside Los Alamos, New Mexico, challenged the structure that the National Park Service had developed. In many ways, the situation at Bandelier was reminiscent of the Waterfalls Canyon fire in Jackson, Wyoming, in 1974, but with an important difference. The nearby Los Alamos Scientific Laboratory (LASL), where the atomic bomb had been developed in the 1940s, remained a primary scientific research facility for the U.S. Department of Energy. Significant scientific laboratories, weapons development facilities, and hazardous materials abounded near the national park on the Pajarito Plateau, about thirty miles due northwest of Santa Fe, New Mexico. If residents of Jackson were disturbed by plumes of smoke in 1974, the health risks from a severe fire near Los Alamos were significantly greater.

A suppression regime had been in force in northern New Mexico since the New Deal. The combination of agriculture, development of the Los Alamos facility, and postwar growth of the region contributed to the policy of suppression of all fires on the plateau. By the 1970s, fuel loads were elevated throughout the area, triggering precisely the kind of situation Kilgore feared when he observed that "in the long run, fuel accumulates and another manager at a later time faces an even tougher decision." The vast increase in density of groundcover and the almost total halt of the natural cycle of ecological replacement that natural fire had long prompted increased the likelihood of a major conflagration at Bandelier or elsewhere in the Jemez Mountains. Testing at LASL compounded the general uneasiness of the people of Los Alamos and the rest of the plateau. The possibility of an accident at the lab igniting a serious fire remained very real.[20]

While federal agencies did an excellent job of suppressing fires on the plateau, the region's fire history suggested that this could not last forever. NPS fire managers experienced dozens of similar situations that preceded difficult fires, and they awaited any outbreak with dread. It arrived late on the afternoon of June 16, 1977, as a spark from a cigarette or a motorcycle engine smoldered in a pile of leaves on the Mesa del Rito in the Santa Fe National Forest. "The fire was started by a couple of kids on motorcycles back up in the woods one day," NPS fire specialist John Lissoway recalled. "The wind was blowing, it was hot, in the middle of June. I think they were out there without a spark arrestor or smoking cigarettes."[21] It grew into the largest fire on the Pajarito Plateau in the twentieth century.

Known as the La Mesa fire, this blaze illustrated the problems of the new NPS fire regime. Human induced, the fire met the conditions for suppression, but it spread so quickly that the response became a valiant effort at containment. Hot, dry, windy weather and dense fuel loads near the ignition point quickly fed the fire. Within ninety minutes of the initial sighting at about 4 p.m., the fire covered more than fifty acres. It spread from the Mesa del Rito area into the national monument by midnight on June 17, and by noon the next day, the fire crossed State Highway 4, headed toward Los Alamos. It grew in concentric circles each day, spreading on the east to within about three miles of Bandelier's headquarters at Frijoles Canyon. Weather exacerbated the situation for the next few days, as winds revived the fire a number of times just as it seemed to lose intensity. By June 21, intermittent thunderstorms slowed the fire, and officials declared it contained in what was conservatively estimated at 15,000 acres at 3 p.m. Two days later, after continuing heavy rain and cool temperatures, the fire was considered to be under control. Only the most vigorous efforts and a complete commitment of resources prevented the fire from reaching the LASL technical areas southeast of Los Alamos, and for at least a day, the town itself was in danger.[22]

The damage was devastating. Raging for more than a week, the fire burned across more than 23,000 acres, including more than 10,000 acres of timber in the northwestern portion of Bandelier and an additional 5,000 acres in the adjacent national forest and on LASL land. The NPS evacuated families at the park headquarters at Frijoles Canyon early in the fire. Cinders and burning ash fell in the nearby town of White Rock, about seven miles from Los Alamos. Wood-shingled roofs there were hosed down constantly to prevent them from igniting.

The fire demanded every available human resource. Firefighting crews from the Los Alamos Scientific Laboratory, the NPS, the Forest Service, and other federal, state, and local agencies joined together to face the threat. The effort enlisted 1,370 people to stop the fire's progress, supported by nine bulldozers, twenty-three ground tankers, five helicopters, and five air tankers. Firefighters swung their Pulaskis—a combination mattock and ax—in twelve-hour shifts around the clock in the heavy smoke; many slumped exhausted at the end of their shifts, to rise again in the morning and repeat the battle.[23]

One of the fire's most dangerous dimensions was its possible interaction with the Los Alamos Scientific Laboratory. The National Park Service had never faced a fire in proximity to a scientific research facility; in the rare cases that an NPS fire had intruded on the activities of the military-industrial complex, Department of Defense and military fire crews had handled the conflagrations. Los Alamos was peculiar. A subsidiary of the University of California system, it lacked the personnel to respond to such a blaze and was forced to rely on the NPS and the Forest Service. Many of its technical facilities were close to the burn area, and due to national security concerns, no one outside of LASL knew what they contained. As the fire penetrated the park and approached LASL facilities, NPS officials had two major worries: LASL was politically powerful and secretive, leading the NPS to act gingerly, and there were real constraints on any strategy. LASL was loaded with combustible and toxic material, and its officials could not provide what the NPS regarded as essential information. Bandelier fire specialist John Lissoway remembered that officials at LASL "did not know how much heat it would take to blow [up]" stored explosives, and the NPS was not privy to the location, quantity, and character of such stockpiles.[24] Managing this fire required even greater political skill and calm than any of its predecessors.

The La Mesa fire became an important test of the multifaceted mission of resource management. Along with the evolution of the use of fire as a tool, fire management included cultural resource management. The burned area was filled with subsurface prehistoric ruins, and quick thinking by NPS officials allowed scrutiny by archaeologists, who preceded the firefighting bulldozers. Although preservation of cultural resources had been an ongoing theme in fires at places such as Mesa Verde, they had never received the attention directed at natural resource management during and in the aftermath of fires. This new level of engagement came about serendipitously. On his way to visit an archaeologist friend at the NPS regional office in Santa Fe, regional scientist Milford R. Fletcher, the head scientist for the NPS in the Southwest, looked up and saw the smoke of the La Mesa fire. He told Cal Cummings, an NPS official responsible for cultural resources, that the situation demanded archaeologists ahead of the construction of fire lines. Archaeologists could locate buried sites and direct the bulldozers away from them, Fletcher insisted. Cummings, Superintendent John D. Hunter of Bandelier, and Santa Fe National Forest supervisor Cristobal Zamora agreed; Cummings found and scheduled volunteers, and Fletcher provided supervision. Nearly forty archaeologists worked in front of the bulldozers during the La Mesa fire, establishing the primacy of cultural resource management even in a particularly dangerous fire.[25]

The fire promoted new cooperation and awareness, but there were tense moments. In one case, Fletcher turned off a USFS bulldozer, telling its driver: "We don't care if the trees burn. They'll grow back. Ruins won't." Although managers made every effort to let archaeologists record sites and guide firefighters away from ruins during the initial construction of fire lines, they often were not present during any subsequent widening. More than 40 percent of the archaeological sites surveyed in the aftermath showed signs of damage. The remainder were unaffected, a measure of the success of the improvised program, and veterans of the fire remembered that shared objectives superseded occasional conflicts. After the fire, archaeology became one component of fire management in the Bandelier area.[26]

La Mesa illustrated the changes in the National Park Service as a fire management organization. It showed that the success of fire management depended on an array of values to guide decisions. The NPS had modest credibility as a fire suppression organization prior to the change in policy reflected in the Green Book in 1968. The NPS considered an overblown suppression organization to be an expensive liability. By La Mesa, other agencies, most notably the USFS, had begun efforts to rein in their fire programs and go in new directions. But as common as a weakened suppression organization became, it did not necessarily assure a strong overall fire management structure. Even as it pioneered fire management, the NPS was seen as strong in theory and rhetoric, but limited in its on-the-ground response. La Mesa illustrated this difficulty with some clarity.

The fire near Bandelier National Monument offered a test to the National Park Service, and the service weathered it. Although the fire was neither prescribed nor natural—it was more typical of suppressed fires in that it resulted from human carelessness—it provided an important challenge to the policy of allowing some fires to burn. La Mesa further illustrated the shortcomings of suppression as a dominant strategy. It did more than all the press releases in the world to remind the public of the danger that fire presented. During a major fire year in the West, with California already aflame, La Mesa was a smaller, specialized blaze that highlighted the NPS's concerns more clearly than it spoke to the larger issues of fire management. La Mesa made the pronouncements of the value of fire sound hollow. The fire garnered publicity and threatened a community; suppression seemed the natural and the only response. It became a reality check for NPS fire managers.

In the aftermath of La Mesa, Bandelier National Monument developed a new fire management plan that took into account the lessons learned on the Pajarito Plateau. The long years of suppression had created a fuel load so heavy and so dry that it altered the composition of the soil beneath and the patterns of regeneration that followed the fire. The new fire plan responded to these realities, clearly recognizing that cyclical burning did more to bring the park closer to the ideal of a pristine environment than did suppression and concomitant catastrophic fire. The objective of the Bandelier fire plan was to "where possible, re-establish the role of fire as a natural process necessary for the perpetuation of fire-dependent ecosystems." The plan proposed prescribed burns for research purposes, keeping the plots small to keep smoke releases minimal. A rigid schedule of conditions under which fire would be allowed was designed, coupled with a careful schedule for preparation of the land and protection of surrounding resources. Such a plan was far from the model of suppression that characterized the plateau through much of the twentieth century.[27]

La Mesa also altered the terrain in which fire and cultural resource management interacted. NPS-18 made the suppression of fires that threatened cultural resources an objective of policy, complicating management at many parks with significant cultural resources. Fuel accumulation and stand density on 338,000 unsurveyed acres in the Grand Canyon created a management problem. It served resource management purposes to let such areas burn if they ignited, yet NPS policy dictated that suppression was in order on land that had not been surveyed for cultural resources. The park faced the dilemma in 1981, seeking authority to allow natural fires within prescription conditions to burn, even if the area had not been surveyed for cultural resources as required under the amended National Historic Preservation Act.[28] Still, even after La Mesa, cultural resources remained a largely unexplored theme in the redefinition of the role of fire in national parks.

Highlighting the problems with prescribed natural fires—letting natural fires burn—the Ouzel fire at Rocky Mountain National Park even more clearly illustrated the gap between ideas about fire management and the realities on the ground. Ouzel began on August 9, 1978, and when National Park Service lookouts discovered the fire a few days later, park officials decided simply to monitor it, in accordance with the park's wildfire management plan. The wildfire plan followed the dictates of the revised Green Book and NPS-18, creating a zone in which fire would be allowed to burn and detailing the conditions under which the NPS would monitor it. It also clearly stated that Rocky Mountain's fire prevention program would "eliminate as completely as possible all man-caused fires," accentuating the difference in response to various kinds of ignition. The 1977 plan defined three zones—low, moderate, and high risk—with different variables to mark them. In the low-risk zone, mostly above 10,000 feet in elevation, lightning fires were to be monitored and allowed to burn; in the moderate-risk zone, below 10,000 feet in elevation but excluding developed zones, natural fires were allowed to burn when the National Fire Danger Rating System index remained under 14. In the high-risk zone, which included the park's developed areas, suppression remained the sole response to fire.[29]

This complex method of response made good sense. By 1977, the National Park Service had developed a complicated vision of fire, combining a burgeoning respect for the value of fire with pragmatic objectives such as the protection of life and property. Even if allowing fire everywhere had been a desired goal, political and cultural constraints made such a strategy unwise at best. The Rocky Mountain National Park wildfire plan served as a model, a delicate but judicious attempt to balance the various forces pulling at NPS fire policy.

For more than a month, the National Park Service limited its response to the high-altitude Ouzel fire to monitoring, with park officials deciding daily whether the fire remained within management objectives. During most of the month, the fire remained within the low-risk zone defined in the fire management plan: the area above 10,000 feet in elevation largely composed of Engelmann spruce–subalpine forests below the timberline and of grasslands, meadows, and rock fields above it. Rocky Mountain staff occasionally introduced suppression tactics, but only to keep the fire within the designated low-risk zone. The ongoing effort to assure that the fire remained consistent with the objectives of park management taxed its resources, but largely avoided public rancor.[30] The sense of a controllable problem changed as high winds swept the park on the afternoon of September 15 and again on September 16. They caused the fire to make substantial runs outside the management zone, threatening the town of Allenspark, Colorado, just beyond the park's border.

The NPS reacted too slowly to the wind change. At about 11:30 a.m. on September 15, Rocky Mountain staff predicted that the fire would escape into the moderate-risk zone. This prospect triggered suppression activity, but before serious efforts could begin, the high winds created a crisis, putting the town in immediate danger. NPS and local response began in earnest, but it was not sufficient to halt the fire's spread. The town was saved by a fluke of geography. A small ridge deflected the chinook wind up and over the town, sparing it. As the fire spread, the National Park Service requested outside help. An elite Department of the Interior team from the Boise Interagency Fire Center headed to the scene. Fire teams were small cadres of "overhead," the staff for the shock troops who do the grunt work on the fire lines. When the team arrived on September 16, professional suppression efforts began with new intensity. More than 600 people battled the fire. High winds dogged suppression efforts for two weeks, until on September 30, 1978, the fire was declared under control. It was finally extinguished on December 4, 1978.[31]

Ouzel was the first time a prescribed natural fire had genuinely threatened a community. Waterfalls Canyon at Grand Teton in 1974 served as a precedent for the National Park Service, but because it was both slow moving and far from the town of Jackson, its only direct impact was the unpleasantness of the smoke. The people of Allenspark felt the real threat of a wildfire that had been permitted to continue to burn by a public agency. As was any such fire, Ouzel was a significant public relations and constituency problem for the National Park Service, reprising an older split between national and local constituencies about the western environment. While *Time* magazine might espouse the NPS perspective on natural prescribed burning, as it had at Waterfalls Canyon in 1974, many people in north-central Colorado perceived only a threat to their homes due to the irresponsibility of a federal agency. In a loud assertion that federal fire management had failed, Boulder County actually fined Rocky Mountain for violation of local air quality statutes.

The Ouzel fire assaulted National Park Service management on two levels. It challenged the orderly structure that the NPS had developed to manage fire, illustrating what the emphasis on process seemed to overlook: fire would not easily conform to administrative dictates. The NPS followed its fire management plan at Rocky Mountain; the results were not what officials intended. In addition, the Ouzel fire cost the NPS the trust of its neighbors and by implication, the trust of the neighbors of any park where fire ran the risk of escaping its human monitors. To those outside the government, NPS policy seemed to place nature above people, a prospect that galled residents of the gateway communities that surround national park areas across the country. Earning back the confidence of local communities was crucial, but it would be an extremely difficult process for the National Park Service.

Pressure from the communities surrounding Rocky Mountain compelled the NPS to explain its choices and to suggest new remedies and strategies even while the fire still burned. On October 3, 1978, a few days after the fire was declared under control, but well before it was extinguished, Superintendent Chester L. Brooks called for a board of review to investigate the fire. NPS official Kenneth Ashley was selected as chair. Herman Ball, a fire management specialist from Region 2 of the Forest Service; Ron Gosnell, the district forester for the Colorado State Forest Service; Richard D. Laven of Colorado State University; and Robert Sellers, an NPS fire specialist at the Boise Interagency Fire Center, comprised the committee. They received three charges: to assess the adequacy of Rocky Mountain's fire plan, especially its provisions for natural fire management; to determine whether the implementation of that plan had been sufficient; and to review the park's suppression efforts once Ouzel was determined to be a wildfire.[32]

With stunning candor, the review board offered an indictment of the application of the existing policy. The park's wildfire management plan provided one target. The board found that deficiencies in the plan "may have conspired to prevent users of the plan from making proper decisions." The reviewers regarded the plan as a statement of philosophy, not an operational directive, exposing a glaring hole in NPS preparation. Rocky Mountain had an exemplary fire plan, written by David Butts, the future head of the Branch of Fire Management. The critique strongly suggested the need for internal rethinking of the park's fire management procedures and practices. It pointed to a lack of information about the park's fire history and an inadequate emphasis on external considerations, such as air quality, adjoining development, and the increasingly urban character of surrounding lands, as causative factors in the park's unfortunate situation. The three concerns encapsulated the history of NPS fire management issues: too few resources, too little scientific information, and a public that did not understand NPS objectives with regard to fire. The review pointed out that Rocky Mountain's plan did not "pinpoint responsibility for decision making," nor did it establish qualifications for the personnel who were to implement the plan. Existing park planning did not contain alternative measures to respond to contingencies such as when fire exceeded a prescription, nor did it include a "precise and separate" action plan. Its criteria for prescriptions to manage natural fire were unclear and insufficient, the reviewers noted. Simply put, the park needed more than Burning Index guides.[33]

The board of review offered a number of ways to improve Rocky Mountain National Park's response in future fire episodes. The report stated that the natural fire management plan should clearly describe contingencies under which suppression would become necessary. The reviewers insisted that the NPS needed to bring the best expertise to the planning process, not reserve it for the aftermath of fires. The development of better fire management units, more clearly delineated by fire history, vegetation types, fuel loading, elevation, and other factors would improve planning and encourage better decision making. The authors advocated considering prescribed fire—the intentional setting of fires—as an additional management tool for the park. More public comment was necessary, not only because of legislation, such as the Resource Conservation and Recovery Act of 1976 (RCRA), which made public input a requirement but also because such interaction built support for NPS programs and created a constituency that would support the service during difficult times.[34]

The report also found fault with the implementation of the park's plan. Although the board of review found that Rocky Mountain's monitoring procedures met NPS standards and functioned well, implementing them proved to be a far more difficult task. Observations of the Ouzel fire were sporadic and incomplete, the review found, and park personnel lacked appropriate information. Spot weather forecasts were not requested in a consistent manner, and as a result, even though meteorologists anticipated the change in conditions, the park did not have sufficient warning about the conditions that erupted on September 15. Fire monitors had not always received clear and comprehensive instructions about their duties. The review board discovered that field notes were almost nonexistent; most monitors relayed information by the airwaves. Radio logs comprised the sole written record. "It appears to the Board that the opportunity to gather important data was lost," the report sternly averred.[35]

The reviewers constructed their own version of the path that took the Ouzel fire out of control. In this iteration, the blaze went beyond the prescription boundaries on September 5, at which time the park's fire committee opted to continue to let the fire burn. This declaration of culpability could easily be regarded as perfect hindsight, but as part of its after-the-fact assessment, the board pointed to a number of factors that contributed to its determination. One part of the fire dipped below 10,000 feet in elevation, entering the moderate-risk zone, where fires were only allowed to burn if the Burning Index were below 14. The higher number of the index that day should have triggered suppression, the reviewers said. The organized local and regional fire response crews that should have been available to Rocky Mountain were busy at other fires, an absence that should have warned park leaders to be cautious. Spotting and crowning combined with the higher Burning Index to create erratic behavior, another trigger for suppression.

The review of the Ouzel fire pinpointed some of the most important problems associated with the new strategies of fire management. First and foremost, funding was essential if parks were to achieve their management goals. One of the review board's most significant criticisms was that Rocky Mountain's plan, wholly adequate as a response to fire, was not appropriately implemented. Between the lines, the reviewers intimated that successful application of the management plan would have prevented the problems that arose. This assessment was simultaneously farsighted and disingenuous. It accurately described a crucial issue that dated from the beginning of the National Park Service, the lack of adequate resources to meet obligations, even as it committed the fundamental and base error of treating wildfire as a bureaucratic category subject to the dictates of a management plan.

At about the same time, Congress added new holdings that transformed not only the national park system, but also its response to fire. As a result of the first serious attempt to adjudicate the land claims of Alaskan natives, the National Park Service acquired what in effect became another national park system in Alaska. The Alaska Native Claims Settlement Act of 1971 (ANCSA) allowed the secretary of the interior to set aside as much as 80 million acres of public land in Alaska for inclusion in federal land reservations. A seven-year dispute ensued, and when no resolution appeared likely prior to the December 18, 1978, date on which withdrawn lands reverted to public domain, President Jimmy Carter proclaimed fifteen new national monuments, eleven of which the NPS was slated to administer. This greatly expanded the NPS's holdings in Alaska. Two years later, the staunchly anti-environmental Ronald Reagan won the 1980 presidential election. Before he took office, Congress offered the nation a lame-duck conservation gift, the Alaskan National Interest Lands Conservation Act (ANILCA). Under its terms, many of the national monuments established in 1978 became national parks or preserves, and the national park system gained more than 51 million acres in Alaska.[36]

The new Alaskan parks presented an enormous challenge for fire managers. The acreage added in 1978 was significantly larger than the entire national park system in the lower forty-eight states and Hawaii. Although the NPS remained focused on the crown jewels of its system—Yellowstone, Yosemite, and their peers—the burned areas in Alaska, Everglades National Park, and Big Cypress National Preserve dwarfed the burned areas in those premier parks. Fire response in Alaska compelled cooperation with federal, state, and native Alaskan entities. The boreal forest burned in an episodic fashion, making it impossible to build up and maintain a large fire response force year after year, simply waiting for the one that brought the big fire season.

Alaska reprised an earlier kind of fire landscape, one in which the nature of fire overwhelmed the human ability to respond. Combined with the dictates of wilderness management, this sense of diminishment—so fundamentally contrary to the ideals of suppression—simultaneously encouraged the practice of allowing prescribed natural fire. Suppression had not been a characteristic feature of the Alaskan landscape as it had in the lower forty-eight states, for the lack of inhabited land and sparse population made the idea of suppression ludicrous.

Suppression in Alaska only really began with statehood in 1959 and developed in the late 1960s. This obviated many of the problems of heavy fuel load that so dogged parks with suppression histories, for natural regimes of fire did not create the density of combustible material that characterized the lower forty-eight. Even more, the size of the new parks guaranteed that fire would be a constant presence. Lightning fires far from human eyes were endemic in the new Alaskan parks. In most instances, these fires burned beyond the reach of park staff. When they were aware of such distant conflagrations, they often lacked the resources to respond.

National park areas had been an important part of Alaskan history throughout the twentieth century, but the National Park Service rarely enjoyed the largesse of resources to devote to its assets in the far north. Only Mount McKinley National Park, later renamed Denali, received substantial funding; other park areas, from Sitka National Monument to Glacier Bay National Monument, languished without comprehensive investment by the NPS. Many national park areas in the state were watched over by volunteer custodians; others were staffed on a seasonal basis.[37] This resulted in a glaring absence of NPS staff in Alaska long after the same condition had been resolved in the rest of the nation.

The NPS needed peer agencies to help with its Alaska parks, and it had to develop new relationships at the same time that it cultivated existing ones. As a parallel Department of the Interior agency, the NPS relationship with the BLM far exceeded any ties to the Forest Service throughout the 1950s. The NPS and the BLM found many areas in which to cooperate. The BLM had the most highly developed fire response system on public lands in the nation's northernmost state, a direct product of its desire to show the firefighting world that it was as competent as the Forest Service.

The shift of lands to the National Park Service in 1978 did not include large sums for their management. Use of the Antiquities Act of 1906, the primary tool available to presidents for the rapid protection of federal land, did not carry the power to allocate funds. Since the Jackson Hole proclamation in 1943, which led to a lawsuit against the U.S. government, presidents had been reticent about invoking the act without prior congressional approval. A tacit agreement between the executive and legislative branches existed; presidents could proclaim any national monuments they wanted, but Congress only would fund the ones it approved in advance. The Carter era national monument proclamations caught the National Park Service in a conundrum. While service officials were pleased to have the new lands, they had to cobble together resources for their management.[38] The vast quantity of land included in the 1978 proclamations forced the NPS to extend its long pattern of reliance on the BLM in Alaska.

The new Alaskan national monuments required the NPS and the BLM to redefine their management arrangements. Although the National Park Service had long relied on the BLM for fire protection in Alaska, the NPS operated under the aegis of NPS-18, while the BLM retained an older suppression standard and remained uncomfortable with policies that encouraged the use of fire. As BLM fire specialist William Adams observed in 1974, the BLM did not have a blueprint for coordinating such activities over the immense spaces of Alaska. Nor did Adams believe that the BLM had enough research to develop a viable program.

The BLM assumed responsibility for fire detection and suppression on NPS lands with the exception of Alaska Railroad and Parks Highway rights-of-way in Mount McKinley National Park. BLM officials agreed to train NPS fire staff if space were available, to assist on NPS prescription burns if the BLM were reimbursed for its costs, to undertake the preliminary investigation of fires where human causes were suspected, and to provide daily situation reports. In return, the National Park Service promised to provide fire prevention programs for national park lands, to rehabilitate its own lands, to report all fires detected on NPS lands to the BLM district office, to collect fire weather data for the parks, and to identify lands that needed protection.[39]

The BLM's growing position in fire management was an asset for the NPS that continued the ongoing transformation of the federal response to fire. It furthered the development of a strong Department of the Interior presence in fire management, which countered the Forest Service. In 1978, a decade after the NPS initiated fire management and with an array of internal struggles over the question, the Forest Service finally embraced the use of fire as of ecological value. The 10 a.m. policy and the parallel ten-acre policy were finally replaced with a program that promoted fire by prescription. By 1978, the revolution in fire practice was complete.[40] Not only had fire management replaced suppression in the Department of the Interior, the Forest Service, where the allegiance to suppression bordered on religion, had finally ratified the new approach.

The pressure for greater cooperation among federal agencies in Alaska grew, in part because the structure to support such a goal was already in place. The National Wildfire Coordinating Group, formed in 1973, provided an avenue for different agencies to work together in a constructive fashion. The Boise Interagency Fire Center (BIFC) offered another avenue for cooperation. In 1978, federal agencies combined in an important experiment in Fortymile, a 12-million-acre section of east-central Alaska. A study team composed of personnel from the National Park Service, BLM, Forest Service, U.S. Fish and Wildlife Service, Bureau of Indian Affairs, Alaska Forestry and Fish and Game departments, and Doyon Regional Corporation, an Alaskan native corporation, assessed the many approaches to fire and assembled a fire management plan. The Fortymile effort was the first of its kind, a harbinger of greater cooperation in the lower forty-eight states as well as in Alaska. "If it can work here, where land plans are as complicated as anyplace in the United States, it can work anywhere," the BLM's Claire Whitlock insisted.[41]

The Fortymile effort created the context for cooperation. David Kellyhouse of the Alaska Department of Fish and Game, the leading proponent of the value of fire for wildlife habitat in the state, championed the Fortymile project and proposed its use as a pilot area for the development of revised management standards for Alaska. A dynamic program that paved the way for future management reform ensued. Even with the state's acceptance of the Fortymile project, the federal agencies had to learn to respect each other's policies and strategies. "I'm sure you appreciate our desire to avoid dividing individual [national] monuments into too many planning units," NPS fire management officer William Paleck reminded Whitlock in September 1979. "We, in turn, appreciate the need to follow natural boundaries and maintain the integrity of fire zones within the state." By early October, the Fortymile fire plan was complete, and the Alaska Land Managers Cooperative Task Force selected two new areas, the Kenai Peninsula and Tanana-Minchumina, as candidates for the immediate development of fire management plans. An interagency public information program was planned as well.[42]

The new relationship was not perfect, for the line between fire suppression and management often was hard to distinguish. In 1980, Paleck notified the Fire Organization Working Group that the service favored a single suppression support organization, but would retain control of fire management planning on NPS lands. David Butts of the Branch of Fire Management summarized the differences that prompted Paleck's plans. The BLM regarded fire suppression as something apart from resource management, a perspective that did not work for the NPS. National Park Service officials in Alaska saw fire management as a complete process that included prevention, presuppression, suppression, and prescribed fire, all in the service of larger resource management goals. By 1981, Butts saw the BLM leaving the NPS out of the decision-making process, and he saw a "high potential for confusing or possibly even contradictory actions" by the BLM. The difference between the two perspectives meant that the NPS had to accept direct management responsibility for a number of functions for which it had long relied on the BLM.[43]

Alaska offered among the most demanding operational fire situations ever faced by the service. As Butts predicted, the service would have to handle prescribed natural fire alone. The amount of work was immense, the demand for resources insatiable, the possibilities frightening, and everything had to be decided immediately. In Alaska, "one of the fun things was the fact that we didn't have time to think," recalled John E. Cook, who served as director of the Alaska area office beginning in 1979 and became regional director on December 2, 1980, when the Alaskan national monuments became national parks and the Alaska Regional Office was created.[44]

Cook understood the pace of work and the need for dramatic and bold action. In 1980, the NPS had assumed responsibility for suppression on its Alaska lands from the BLM, but found itself unprepared for the responsibility. Under the arrangement, the BLM agreed to provide basic suppression services for the immediate future. The NPS was to provide the land manager's representative (LMR), which Cook described as the surrogate for the superintendent in fire situations. The fire boss of any specific conflagration would report to the LMR. "This is an important step which can not be delayed due to the breadth and scope of the environmental and economic impacts of fire suppression within the State as well as changing agency roles and relationships in Alaska," he informed other regional directors. "We need your help."[45]

Cook's dilemma was simultaneously simple but insoluble. In Alaska, the NPS lacked enough people who could serve as LMRs in the case of a significant fire year. The BLM, already in transition as a result of the fire circumstances of Alaska, was generous in its willingness to support the NPS; Cook needed to be able to match its peer agency's support with NPS resources. "I am asking the Regional Directors to assist us by providing the nomination of any qualified individuals for detail assignments as Land Manager's Representative," Cook beseeched his colleagues. "No one looks forward to the day when we in Alaska can supply as much assistance to other regions as we have received, more than I."[46]

With the debut of the BLM Level I draft plan in March 1981, the BLM emphasis shifted toward suppression on the newly designated native lands. This change, and the ongoing focus on suppression as NPS fire managers pursued different paths, limited the BLM's effectiveness for the broader-based NPS management policy. By October 1981, differences had overwhelmed the cooperative ethos, and the relationship had crumbled over the wording of a BLM departmental manual. Each time NPS officials felt they had acceptable language, the BLM offered further revisions. The situation was "time consuming and frustrating on the part of both of these staffs," Butts asserted. "The role of Alaska BLM to provide logistical support, retardant aircraft, smokejumpers, etc., is not challenged by us," he continued. "But the Bureau of Land Management can not and is not in a position to provide monitoring of prescribed natural fires" that occurred on national park lands. The BLM had sought to prove it was as good at suppression and firefighting as the Forest Service, so it had adopted a hardcore suppression approach based heavily on smoke jumping. This did not last, but it complicated discussions between the two agencies. The BLM behaved like the old Forest Service, and Butts felt that the situation intruded on the authority of the National Park Service. If the BLM handled fire suppression on Fish and Wildlife Service lands as well as on native lands, the pressure on the NPS to allow suppression would be enormous. He proposed maintaining the ongoing suppression arrangement, but writing a revised "fire management program, which will be the primary tool in resource management for Alaskan natural area parks." This translated into a different vision of policy: "The National Park Service does not intend duplicating BLM suppression capabilities or forces, but does intend to complement them in order to accomplish full spectrum fire management programs within the national parks."[47] Different in its needs, the NPS decided it would have to go it alone—with all the responsibility that departure from the cooperative arrangement entailed.

Pressure from the highest levels of the Department of the Interior helped the National Park Service to clarify its position and responsibilities. NPS director Russell Dickenson strongly and successfully argued for an articulation of the differences in the service's mission. Dickenson persuaded the assistant secretary of the interior for fish, wildlife, and parks, G. Ray Arnett, to advance the NPS perspective. Arnett informed his counterpart, the assistant secretary of the interior for land and water resources, Garrey E. Carruthers, that the Department of the Interior "should pursue a course of action that accommodates the necessary variation among the bureaus as long as they are not redundant." During the Reagan administration, under Secretary of the Interior James Watt—who challenged conventional conservation at every opportunity and promised the press that he would "use the budget system to be the excuse to make major policy decisions" that strangled programs he did not like—this stance reflected a broader vision of the NPS mission than was typical among senior Department of the Interior officials at the time.[48]

A new interagency agreement quickly resulted. The agencies formalized a new policy over the winter of 1982, before the summer fire season started. Under it, the BLM's role changed dramatically. It relinquished administrative responsibility for more than 200 million acres of Alaska but retained its primary leadership role in fire suppression even as those lands were turned over to the state of Alaska, Alaskan native corporations, and Department of the Interior agencies. The result was a forced compromise, essential to management of the far north. The NPS entered into a fire management program that became the "primary tool in resource management for Alaskan natural area parks." The BLM established the Alaska Interagency Fire Command in Fairbanks as the central fire response facility. This sole statewide fire suppression organization served as the "initial strike force against wildfires" on almost 300 million acres of Department of the Interior lands. The BLM's arrangement with the NPS was formalized with a new interagency agreement in May 1982.[49] Even in an era when the secretary of the interior was an unabashed opponent of conservation, fire was too threatening and its management too important to be left in chaos to hew to the antifederal line common in the Reagan administration.

The NPS's resources for fire suppression remained vastly limited in comparison with the BLM, and to earn credibility, the service had to contribute to the national pool from which it drew so often. In 1981, the NPS's Office of Fire Management announced a pilot program to create three crews to "assist all land managers with their fire problems." Known as Arrowhead No. 1, No. 2, and No. 3, the crews were designed to meet the specifications for full-service Class I teams. The NPS crews were composed of nineteen people, including a crew boss and three squad bosses. They were expected to provide support for the initial response of parks and to contribute the NPS's share of the interagency fire crews.[50]

This was the situation in which Brad Cella found himself when he arrived at Wrangell–St. Elias National Park in 1982. A veteran of Yosemite and the NPS resource management training program, Cella said he was "shaking my head saying, 'I know they could have got someone better than me to be the first resource manager at the largest national park in the nation.'" When he arrived, region- and areawide fire planning in Alaska had just begun. William Paleck, who had formerly been the regional fire management officer, had become chief ranger at the park, so Cella was reporting to one of the most experienced fire management people in Alaska. Paleck "was willing to let me run with fire because he could watch what I was doing," Cella recalled, and he became the National Park Service representative to the Copper Basin fire planning effort.[51]

Fire management in Alaska evolved into the most integrated and comprehensive interagency cooperation in federal land management. When representatives of the land management agencies sat down to discuss their options, each proposed his or her agency's vision of the situation. Then the negotiations began. "The attempt was to try to ignore agency boundaries and look at the fire environment and look at the values to be protected," Cella recalled. "I think the absolute key was that we talked about values, not each other's values. I didn't try to tell the Forest Service what was important to them or the BLM what was important to them. And they, by and large, didn't try to tell me what was important to the National Park Service." The negotiations focused on "how we could draw a line on a map," Cella observed, "but it wasn't over our values. I think it really kept us out of a lot of sticky stuff."[52]

The combination of agency programs and experience yielded significant results at all levels. The National Park Service became an important component of the Boise Interagency Fire Center, which became the National Interagency Fire Center (NIFC) in 1993. In no small part, the intensive cooperation that led to the NIFC grew out of the cooperative experience of Alaska. Long a debtor to other federal agencies when it came to fire resources, the NPS became a significant contributor to interagency fire efforts. In 1986, the service participated in 159 mutual aid dispatches, in which 28,761 acres of other agencies' land burned. The NPS participated in the national mobilization of firefighters in August 1986, the second year in a row that such action was necessary. A team of 528 NPS firefighters and staff were dispatched to western fires, and engines from the Western Region and a helicopter from the Rocky Mountain Region contributed to suppression efforts. In Alaska, an NPS aircraft played an integral role in interagency suppression efforts. In turn, several NPS fires also required outside assistance. Five "project fires," as such conflagrations were labeled, required 1,050 firefighters and staff from other agencies as well as the use of twenty aircraft. The NPS had a net gain in 1986. It received more help from other agencies than it provided even during the mobilization in August.[53]

Throughout the NPS, the goals of fire management were implemented in a systematic fashion. The change was palpable; from 224 acres in 10 prescribed burns in 1977, the National Park Service engaged in 108 burns that covered 36,024 acres in 1986. Wildfires remained more random. The years 1981 and 1986 were brutal, with fires covering 95,055 acres in 1981 and 119,976 acres in 1986, but they were aberrations. According to the NPS, a more typical year saw wildfires burning around 20,000 acres. More telling, 145 prescribed natural fires covered 75,491 acres in 1986, but this resulted in no small part from the increased fire throughout the park system that year. More typical was a prescribed natural fire total annual burn in the 20,000-acre range.[54]

1986 also served as a harbinger of a more dangerous and difficult future. Around 1985, what has become a twenty-year drought cycle began, interrupted by a wet period between 1989 and 1992. From the mid-1980s, federal agencies had to impose their policies against the pressure of the long drought. This confluence provided a partial explanation for why more has not happened. 1986 became the worst in National Park Service fire history; the 195,467 acres that burned in wildfires and prescribed burns was the highest total in recorded NPS fire history. This total came on the heels of a difficult previous year, in which fires burned across more than 2.8 million acres of public land throughout the country.[55] Although the fires' impact on the National Park Service in 1985 had been muted, the overall trend and the increasing interdependency of interagency fire response gave NPS fire personnel concern about the future.

The following year was even worse. The 1987 fire season required the largest mobilization of personnel and resources to fight fire in recorded history. Every federal agency in the West contributed a higher level of resources than ever before. Nearly 2.5 million acres burned in 71,300 fires nationally. The NPS experienced a heavy year as well, with 704 wildfires suppressed after burning on almost 39,000 acres. Prescribed natural fires were also significant; 129 such fires burned 12,761 acres. The NPS continued its prescribed fire program as well, with 111 prescribed fires burning on 28,893 acres. During the first half of the year, fires in the Southeast and Southwest confronted the NPS, but the greatest demand on service resources followed outbreaks of fire at the end of August. As California and Oregon burned—in one California fire, 580,000 acres burned in less than two weeks—the NPS contributed to interagency efforts. More NPS fire personnel assisted other agencies in 1987 than in any previous year. The service dispatched more than 1,100 NPS firefighters to the West Coast blazes, also contributing to firefighting efforts in Washington and Idaho and taking all kinds of labor from their home parks. The system required trade-offs and had serious long-term costs.[56]

The fires forced nearby national parks to respond with emergency measures. Near the worst of the Oregon fires in the Siskiyou National Forest, Oregon Caves National Monument readied evacuation plans; at Yosemite, during the Labor Day holiday, one of the busiest weekends of the year, the fires in the Stanislaus National Forest spread into the northwestern part of the park, threatening the Merced and Tuolumne groves of giant sequoias and the nearby communities of Hodgdon Meadows, Crane Flat, and El Portal. The park closed roads and campgrounds as a precaution, and some NPS employees were evacuated from the communities.[57] Such disruption was uncommon but not unprecedented. It further underscored the ever-present threat of fire to the national park system.

Threats from fires outside national parks posed significant management problems at Sequoia–Kings Canyon, but in one major instance, earlier prescribed burning obviated what otherwise might have been dire consequences. The Pierce fire, which started in the Sequoia National Forest, showed extreme behavior, crowning and burning giant sequoias outside the park's boundary. When it swept into the park, into a section of the Redwood Mountain Grove, the scene of one of the first prescribed burning programs in the system, the reduced fuel load could not sustain the fire, and it was controlled with hand lines. The park's sequoias were not damaged, solid evidence of the efficacy and long-term value of prescribed burn programs.[58]

The 1987 fire season further illustrated one of the ongoing problems of fire management. The transition to using fire to control fire had not happened quickly enough. Much of the land touched by fire had been subject to suppression for a long time, analysts recognized, creating the conditions that caused the worst fires. NPS fire personnel could take heart; the most severe and the most dangerous fires were not on NPS lands. It was easy to embrace the limited burning programs on NPS lands and point to them as proof of the success of the theory of controlled burning. Yet millions of acres that had been subjected to suppression remained adjacent to or near national park lands, dramatically increasing the threat to national park lands throughout the country. The years between 1985 and 1987 suggested that the bill for suppression was coming due. In each successive year, fires worsened, and managers viewed the situation with growing trepidation. From their perspective, the successes were small in scale, the threats enormous and growing. Even worse, the faith in prescribed burning and prescribed natural burning had not been matched by action, and NPS lands themselves contained millions of acres that had been subject to suppression for a long time and had not yet been reached by fire management efforts. While the tendency was to regard such lands as one of the consequences of the huge expansion of Alaskan national park lands, the problem was more widespread.

By the late 1980s, a tremendous amount had been accomplished. The decade since the implementation of NPS-18 had been revolutionary. "Fire control," the overarching philosophy of suppression for its own sake, had been eliminated, replaced with an infinitely more sophisticated balance of the use of fire, its introduction, and suppression. The NPS had weathered disasters such as the Ouzel fire, which conversely strengthened management by leading to the establishment of the Branch of Fire Management. After all the confusion of a decade of transition, the NPS had much to its credit. It was becoming professional in fire management, and as the USFS continued to slide, the NPS rose to the forefront among the federal agencies.

Yet the premises upon which these changes hinged were subject to challenge by fire itself. The fantasy that fire was simple and that planning, science, and organization could bring it to heel had been shattered in reality, but not yet accepted on the ground. The destruction of the ideal had begun at Ouzel, but the lesson did not take very well. The emphasis on science and planning, two important watchwords in the post-1960s National Park Service, made experienced professionals less cautious about the realities they knew than they could have been. Fire planning blossomed but, without a comprehensive review process, varied in quality. The best park fire plans were remarkable for their clarity and depth, their foresight and comprehensiveness. Others remained idiosyncratic, strongly reflecting local sensibilities but not comprehending, much less achieving, national objectives. The successes of the decade—the interagency cooperation, a nomenclature change that reflected the growing interdependence of Department of the Interior agencies and their independence from the Forest Service—signaled notable transformations.

Fire management remained an uneven proposition in the NPS. Yosemite, Sequoia, Kings Canyon, Glacier, and the Everglades led the way. Differences in management policy did not keep the new parks in Alaska from the forefront of interagency cooperation. Yet despite its long fire history, Yellowstone did not stand in the front rank of fire management planning. As the spring of 1988 approached, the nation's first national park had an approved fire plan that dated from 1972 and that reflected the concerns of that era. This seemed innocuous, but it proved to be ominous, a portent of an explosion of nature and an implosion of policy that would rock the foundations on which fire management in the National Park Service rested.

SEVEN Yellowstone and the Politics of Disaster

In the summer of 1988, Yellowstone National Park, the most iconic symbol of the American desire to affiliate national destiny and identity with nature, erupted in flames. Americans love Yellowstone as both symbol and actual place. They expect their National Park Service to protect it, especially from something as elemental as fire. Throughout the summer, as the park burned, the American people wondered what was going on. How could such a calamity take place? Who was responsible?

Professional fire planners and managers believed that by adhering to scientific principles derived from research, they could create a system that controlled fire and even turned it to the NPS's advantage. The belief was reasonable, but it failed to take into account the once-in-a-generation event that could not be planned for. The Yellowstone fire was that event: a giant fire in a place so important to Americans that it shattered the fire management program as it had been conceived, illustrating not only the boundaries inherent in the implementation of policy, but the fundamental impossibility that existing strategies could meet the extraordinary challenge.

In essence, conflagrations such as the ones that occurred at Yellowstone in 1988 transformed fire policy from a science-based response to a political issue. As long as fire remained a threat but did not present an immediate and insurmountable danger, scientists and park managers controlled the terms of debate. They could frame the underlying science in practical and abstract forms to buttress their arguments for policy implementation. Against such a carefully reasoned, science-based strategy, those who opposed NPS fire policy sounded shrill, unreasonable, and self-interested. Under such circumstances, professionals had the upper hand supported by the growing body of research that seemed to illustrate the value of fire management.

But the convergence of events in 1988 challenged the entire fire management model of the National Park Service as well as its administration of the parks themselves. In the summer of 1988, more than 1,427,902 acres in the greater Yellowstone area burned over almost four months. That total included 793,880 acres in Yellowstone itself, almost one-third of the park. When a November snowfall finally put an end to the blazes, the nation's first park, symbol for many of the country's relationship to nature and its wisdom in preserving even a small part of it, had burned uncontrollably. With that fire, the National Park Service found its image seriously tarnished by the public's sense of betrayal. The mission of the National Park Service was to protect nature; the devastation that the public saw in the fires of Yellowstone seemed to belie their trust. If the National Park Service earned its stature in fire management in the California parks, it found the limits of its knowledge, experience, and resource base at Yellowstone.

The NPS long had been the most beloved federal agency, providing park visitors with their most positive encounters with the face of national authority.[1] Fire management in general had caused some friction with the public, leading to diminishing loyalty to the service in some quarters, but the public still generally beamed when it looked at the national parks, and it retained real fondness for the people who protected these treasures. The Yellowstone fires accelerated existing tensions and added new dimensions that led to outright condemnation by the media and the public of the NPS, its policies, and even individuals in the service.

The summer of 1988 was the driest on record at Yellowstone National Park. Although the spring had been wet, with 155 percent of normal rainfall in April and 181 percent of normal amounts recorded in May, very little precipitation fell in the park during June, July, or August. Early in the summer, when Yellowstone was still wet, park staff elected to let about twenty lightning fires burn in accordance with policy. Each fire was evaluated on its own merits, the decision to monitor or suppress dependent on conditions.[2] As the summer progressed, conditions for fire to start and spread became common, and the National Park Service and every other land management agency in the region—at federal, state, and local levels—were prepared for the eventuality. NPS officials at the park and the regional office carefully monitored Yellowstone's situation, making decisions based on constantly changing circumstances.

In early June, the situation became threatening, but the risk appeared to fall within acceptable parameters. Fire managers had no reason to believe that any fires that occurred during the summer could not be controlled. Even though the region quickly dried out and rainfall appeared unlikely in the short term, the overall year had been wet to date and the weather pattern of recent years suggested that summer rainfall would soon follow. Fire managers had dealt with very difficult summers in each of the three previous years, handling record levels of fire on federal lands. Confidence ran high among fire managers throughout the federal land management system; prescribed natural burning and prescribed burning had lowered fuel loads where implementation had taken place, and plans for more comprehensive introduction of fire permeated the National Park System. Interagency cooperation modeled on Alaska had taken root at the BIFC in Idaho, and programs such as FIREPRO in the NPS and equivalent programs in other agencies inspired a level of confidence in planning and the deployment of fire resources that had not been possible a decade before. Yellowstone superintendent Robert Barbee, who had come to the park in 1983, was an old fire hand with experience that dated back to the introduction of prescribed fire in the park system in 1968.[3] Fire was always a tough opponent, but in 1988, most federal land managers believed that the tools they had to manage and combat it were equal to the task.

The Yellowstone region began to burn on June 14, when lightning started a fire in the Custer National Forest, north of the Cooke City, Montana, entrance to the northeastern corner of Yellowstone National Park. Called the Storm Creek fire, it began in Absaroka-Beartooth Wilderness and eventually spread over 95,000 acres. New fires continued to start, most induced by lightning. On June 23, lightning struck near Shoshone Lake, a remote area about ten miles from Grant Village. The initial blaze was small, about 70 acres. On June 25, another fire began in the northwestern corner of the park about thirty-one miles west of the north entrance. On July 1, yet another fire ignited east of Yellowstone's southern entrance. The fires multiplied, with new ones ignited on July 5 and July 9.[4] A management nightmare for the National Park Service had begun. Natural fires proliferated, and the NPS had to make quick decisions.

The service initially remained committed to its complicated mix of allowing some fires to burn, suppressing others, and in some cases, initiating prescribed burns in well-defined areas for management purposes. Yellowstone's fire plans remained rooted in the philosophical statements of the early 1970s; some in the park fiercely resisted the more sophisticated programs that had begun in the 1980s. A plan that had been drafted three years before had not yet been sent through the approval process. It would permit as many lightning-started fires as possible to burn; protect human life and property, natural features, endangered species, and historic and cultural sites from damage or destruction; suppress wildfire in a safe and cost-effective fashion; and utilize prescribed burning to reduce fuel loads. Between 1972 and 1986, fires had burned across 34,175 acres in Yellowstone under the prescriptions that allowed natural fire. The largest single burn during that time was about 7,400 acres. The largest natural burn in the park's history, at Heart Lake in 1931, had been only 18,000 acres. Given the scope and scale of NPS experience, the service's actions when the fires started followed policy and reflected the predispositions of NPS experience with fire.[5]

Park managers viewed the early fires in 1988 through the lens of recent experience. In the 1980s, Yellowstone experienced a series of abnormally wet summers. Only once between 1977 and 1987 did the park fail to receive average July rainfall. In four of the five years beginning in 1983, the park experienced more than twice the average monthly rainfall for July. In 1987, the most anomalous year, Yellowstone received three times the annual average rainfall in July. With six consecutive years of above average rainfall in July, park managers and fire behavior specialists decided to continue established practice with what they defined as a natural prescribed fire and simply to monitor the lightning fires.[6]

But 1988 broke with recent history and eventually the shortfall of rain in June and July led to dangerous conditions. During June, the park recorded only 20 percent of the average rainfall for the month; July reached 79 percent of the monthly average. Moisture content in Yellowstone fell precipitously. By the end of July, fuel moisture levels in plants and tree branches were at astonishing lows. In grasses and small branches, moisture levels had dropped to as low as 2–3 percent, well beneath the 15 percent that signaled danger. Dead trees were measured at 7 percent moisture. NPS records showed that when timber was between 8 and 12 percent moisture, lightning served as an effective ignition for fires that burned freely. Even worse, unusually high winds associated with the dry fronts passing through the region spread any flames widely, much more than would have occurred as a result of the dryness alone.[7]

The result was a rapid change in policy that elevated suppression to the primary response in Yellowstone. On July 15, the park no longer allowed new natural fires to burn. When the decision was made, fires inside the park topped 8,600 acres. By July 21, fires covered 17,000 acres, prompting an even more aggressive response. As of that date, every fire in the park was to be fought. An extensive interagency fire response effort began in mid-July. Experienced firefighters found that the combination of extreme weather and dense and dry fuel loads posed conditions rarely encountered. Conventional firefighting techniques, such as burning to create fuel breaks and backfiring, proved to be ineffective. New fires started when winds blew embers from the tops of enormously high trees far ahead of the main fire—and almost always beyond a fuel break or a backfire—thwarting most efforts to contain the fires. Called spotting, this phenomenon made ineffectual even the widest of bulldozer lines. Fires started by spotting crossed the Grand Canyon of the Yellowstone River and routinely jumped roads and streams. As a result, the speed with which the fires moved was stunning. In many instances, fires traveled between five and ten miles per day, with instances of a two-mile jump in one hour not uncommon. This contrasted with norms of closer to one mile per day. The tremendous heat generated by the huge fires contributed to their spread, for it let the fires consume even the heaviest of fuels, which would not have been likely to burn in a normal fire season. Everything about the Yellowstone fires seemed designed to demonstrate that fire could exceed human control.[8]

Secretary of the Interior Donald Hodel toured the area on July 27, confirming suppression as the service's primary objective in battling Yellowstone's fires and reminding everyone that the natural fire program had been suspended. The public and congressional representatives expected to see the results of suppression, to see fires extinguished, and to watch as the dramatic fires of 1988 came to an end. Such a result was simply beyond human capability. Firefighters could not attack the fires from the front, as spotting and the high winds made the risk too great to bear. Crews could be overrun or trapped between the spot fires out front and the main fire behind. As a result, firefighting took place from the flanks except when lives or property were in the direct path of an oncoming fire.[9]

Experienced firefighters were shocked at the fires' power and at the ineffectiveness of all responses. Even those with as many as twenty years in fire response had never seen anything like Yellowstone in 1988. Most agreed that the only solution to fires of this magnitude was help from the weather. Rain or snow could alleviate the conditions, but no technology, strategy, or amount of labor could overcome the flames. "We threw everything at that fire from Day One," observed Denny Bungarz, a USFS incident commander from the Mendocino National Forest in California, who fought the robust North Fork fire. "We tried everything we knew of or could think of, and that fire kicked our ass from one end of the park to the other." Bungarz's sentiments reflected not only the magnitude of the problem, but also the way in which this fire shattered expectations about fire management.

Throughout the grueling months of the fire, the commitment of fire crews and their professionalism exceeded even the highest expectations. Because of the pressure and danger in the work, crews turned over with great frequency.[10] "[T]here was a guy named Dave Poncin who was an incident commander Type I, who was just beyond outstanding. So was his whole team," Superintendent Robert Barbee remembered. "When you lose somebody like that, you really feel the loss." Barbee felt the same toward

Richard T. (Rick) Gale, who served as the unified area commander later in the fire. "He was a star in my opinion," Barbee recalled. "There is a guy who is smart, whose synapses fired cleanly, no carbon build-up. He did a wonderful job." The turnovers led to changes at about the time the working relationships coalesced. "Then you get a complete change and it is disruptive," Barbee insisted. "No question in my mind. Now, I don't know what you do about it, because you can't have those guys in harm's way all the time. They get [too] tired."[11] With mandated turnover in personnel from the highest levels of the agency, continuity was hard to achieve.

New fires continued to start across Yellowstone, while existing separate fires joined together to create even more dangerous, powerful, and threatening conglomerates. By August 2, the Clover-Mist fire topped 73,754 acres as it spread into the heavily timbered Shoshone National Forest. On August 10, the more than 20,000-acre Red fire joined with the 25,200-acre Shoshone fire. Burning in the southern end of the park, the Red Shoshone fire grew rapidly, burning across another 10,000 acres over the next five days. Other fires continued to spread, with the Clover-Mist fire reaching 95,000 acres on August 14 and the North Fork fire at 52,960 on the same day. On August 20, called "Black Saturday," new records were set, with fires burning over 165,000 acres of timber, the highest daily total ever recorded at Yellowstone. "Giant mushroom clouds rose into the atmosphere," observed reporter Rocky Barker, "making it seem like the park was under nuclear attack." Silver Gate and Cooke City, two of the northeastern gateway communities to Yellowstone, soon were in danger. The fire exploded in response to dry, cold weather fronts that produced winds as high as sixty miles per hour. A backburn reduced fuel loads enough to keep the fire from the two towns, but the situation was serious enough that someone added a letter to the Cooke City sign and made the town "Cooked City."[12] It was a fitting modification, given the difficulty of containing the blaze. Still, saving the two towns affirmed the confidence that had been the hallmark of interagency fire management.[13]

The national policy response to the fires was rapid but symbolic. On August 23, 1988, in the midst of the Yellowstone fires, NPS director William Penn Mott declared a freeze on all prescribed burns in the national park system.[14] Mott's decision was a throwback to an earlier era. The suppression order introduced at Yellowstone a month before became a systemwide standard for the first time in twenty years. While such a decision revealed elements of clear and precise after-the-fact decision making, it also demonstrated a heightened sensitivity to public criticism of the service and its practices. Even while firefighting efforts continued, the NPS had returned to trying to prove its worth as a scientific manager and as a steward of public resources.

On September 7, high winds brought the North Fork fire to the Old Faithful complex, the first time fire had threatened the area in the 116-year history of the park. An aerial suppression assault attempted to slow the fire's progress, but those efforts failed. Early in the morning, the National Park Service evacuated the complex. Between 500 and 600 people left by the 10 a.m. deadline, although visitors traveling by car still were allowed to visit the geyser as late as mid-afternoon, some arriving just minutes before the firestorm struck. The fire eventually encircled the Old Faithful area, and firefighters successfully battled to save the Old Faithful Inn and the electrical substation nearby. The fire burned so hot that it melted the rubber off the wheels of cars and a truck, shattered vehicles' windshields, and scorched their paint. As many as nineteen buildings in the area burned to the ground, and the old dormitory building suffered damage. No one was hurt in defense of Old Faithful, although two deaths were associated with the North Fork fire in the greater Yellowstone area.

The North Fork fire was the classic fire that the National Park Service had always combated, resulting from human carelessness. It began on July 22 in the Targhee National Forest, the result of a cigarette dropped into dry leaves by one of four woodcutters who were taking a smoking break.[15] The National Park Service responded with its full capability. But while the fire was typical of those the NPS and other federal agencies had aggressively battled over the years, the conditions under which it occurred were rare. Weather conditions, including high winds and a lack of precipitation, made the situation volatile.[16] In the end, the North Fork fire burned across more than 56,000 acres on September 7.[17]

The fire's out-of-control spiral rightly frightened park staff. Chief ranger Dan Sholly recalled that "not so many weeks ago, I thought the 4,700-acre fire sweeping toward the Calfee Creek cabin was a major blaze. What was it now? I looked at the fire summaries. It was the first one listed: Clover-Mist fire—238,000 acres." Fire again proved to be more powerful than even the most professional planning and modeling, destroying all the assumptions that specialists had made about its behavior. Park ecologist Don Despain had played an instrumental role in designing Yellowstone's natural fire policy and earlier in the summer had predicted that the fires would grow no larger than 40,000 acres. As they approached 1 million acres, he evacuated his family from the park. Despain's research had been the standard on which most modeling had been based, and following his data, leading fire behaviorists predicted that any fire in Yellowstone would consume the available fuel or be doused by rain before August ended.[18] Once again, fire proved that its behavior defies prediction.

The Yellowstone fires were the worst in a year that saw brutal fires throughout the West and Alaska. More than 72,000 fires were reported on federal lands in twenty-two states; 299 of these were classified as major. This designation meant that more than 300 acres burned or that Type I or Type II incident management teams were dispatched, the highest ranking teams in the fire management system. Ultimately, fire burned across more than 4.3 million acres, enhancing the sense of apocalypse that was widespread in the summer and fall of 1988. The BIFC dispatched more than 41,000 fire personnel, including 4,000 temporary firefighters, in response. Between the middle of July and late September, 35,000 people actively fought fires. Almost 6,000 soldiers were deployed. The bills for fighting these fires were staggering. The USFS spent $384.3 million, while the Department of the Interior reportedly added $200 million to the total. The final count showed that the federal government spent more than $600 million fighting fires throughout the region in 1988.[19]

An assessment of the impact of the Yellowstone fires revealed stunning consequences for the park and its environs. Fires raged across more than 1.4 million acres in the greater Yellowstone area; funds in excess of $120 million were spent on firefighting and management. Almost one-third of the burned acreage, 566,608 acres, was inside the Targhee, Custer, Gallatin, Bridger-Teton, and Shoshone national forests surrounding the park. The rest, slightly less than 1 million acres, was inside Yellowstone.[20] This total, nearly 45 percent of the park's 2.2 million acres, represented the most visible evidence of the fires' power and the fundamental ineffectiveness of all human countermeasures. The nation's leading tourist paradise lay in smoldering ruins, with even Old Faithful burned over. It was hard to imagine a more striking symbolic landscape than the visible ruins of the nation's first national park.

The outcry about the NPS response started in August. The media became a constant presence at Yellowstone. The Yellowstone fires were shown live on network television every night, and newspapers across the country featured the story daily. The Cable News Network (CNN), barely a decade old and always in search of anything it could call news, showed hourly pictures of the flames and constantly reported on the growing threat in the breathless manner of television. "It was an incredible episode," Superintendent Robert Barbee remembered. "I kept waiting for Gaddafi or somebody to do something outrageous, because we were the only [news] game in town all summer long."[21]

The national spotlight focused on Yellowstone never wavered. With realities such as the ranch owned by NBC News anchor Tom Brokaw sitting just north of the park, Yellowstone found itself with even greater attention than the crown jewel of the national park system ordinarily merited. Brokaw only changed his view of NPS actions when Old Faithful was threatened in September, telling a national television audience that "there are a lot of angry people who believe that the National Park Service is responsible and has let the fires burn too freely for too long." While Brokaw was only reflecting public sentiment, "we were not really prepared for that kind of media triage [*sic*]," Barbee said in a candid assessment. Media coverage of the event was "superficial and stereotypical," observed Ohio State University journalism professor Conrad Smith, who studied the press response to the fire. He believed that urban reporters brought a set of preconceptions derived from city structure fires that colored their perception of the Yellowstone fires. The media's cameras shaped the view of the Yellowstone fires, contributing to their political consequences.[22]

Barbee found himself at the center of a maelstrom. "I personally became a lightning rod," he grimaced. "By August, it was beginning to take a bit of a toll on me," he recalled. His superiors "kept saying 'well gee, maybe we ought to let someone else come in, and let you take a breather.' And I said no. I argued strongly against that; it would have caused all sorts of problems." Barbee had become what he described as the agent provocateur, the focal point of animosity about the fires. "The worst thing that could have happened would have been for me to step back, and them to bring somebody else in, some other senior person to take over," he insisted. "It would have sent all kinds of bad signals." Abdication would have conveyed a message that the park was admitting that it had done something wrong. Barbee believed strongly that, as superintendent, he should weather the storm of anger and questioning that accompanied the fire.[23]

National Park Service director William Penn Mott sought to help Barbee by explaining the NPS position and its mission. Almost three weeks after the NPS declared that it would suppress all fires in Yellowstone, Mott informed Senator Malcolm Wallop (R-Wyo.) of the service's fire planning objectives. "The flexibility to suppress naturally ignited fires when conditions become extreme, or facilities and adjacent land are threatened is unequivocally part of our policy," Mott assured Wallop. He attributed the difficult fire situation at the time to a combination of high fuel loads and dry weather. "I am pleased to report that with the help of some 2,000-plus fire fighters and professional staff, all Yellowstone area fires are under control," Mott trumpeted a little prematurely on August 11. "Unless extreme weather, such as continuous high winds, occurs, we expect them to remain so."[24]

This letter was identical to ones sent to the governors, U.S. senators, and congressional representatives from Wyoming, Montana, and Idaho.

All three states relied on tourism and the dollars generated by Yellowstone National Park, giving each a vested interest in the NPS fire response. While Mott attempted to persuade each state's political leadership that there was "a positive and pragmatic side of the fires we see today," his argument fell on unsympathetic ears. No matter how he couched the fires—as a "rebirth" or a "renewal of the park ecosystems"—leaders of states that depended on visitors did not accept the service's argument.[25] In their view, the fires were a short-term economic and ecological disaster. After watching Yellowstone burning every night on the evening news, most tourists decided to travel somewhere else that summer, costing every state around Yellowstone enormous revenue.

By early September, the cries against what was perceived as a defective NPS policy reached a crescendo. Even though the service had reverted to suppression in mid-July, a collection of western congressional representatives and senators, mainly Republicans, approached President Ronald Reagan in protest. "We strongly feel the National Park Service policy of 'let it burn' is wrong, especially with the drought and weather conditions in the west," stated a petition by Representative Ron Marlenee of Montana that also was signed by Don Young of Alaska, Jim Hansen of Utah, Larry E. Craig of Idaho, Bob Dornan of California, and Byron L. Dorgan of North Dakota, the lone Democrat to sign. "Ask anyone from the area and they will tell you that this is the wrong time and the wrong year to let a fire burn. The National Park Service did not heed these signs or the advice from many sources of the gravity of this year's fire conditions," the petition charged. The representatives demanded a change in what they inaccurately perceived to be the NPS policy of allowing fires to burn.[26]

This accusatory stance was consistent with the negative feelings such representatives held toward the NPS. Most were "sagebrush rebels" from the decade before, vocal proponents for the transfer of federal land to the states. Many had bought into the larger vision of the Wise Use movement, an appropriation of Gifford Pinchot's language of the idea of the greatest good for the greatest number for the longest time, but perverted for the use of those who saw the consumption of public lands as the singular and sole purpose of them. Despite a changing regional and world economy, a new and overwhelming emphasis on outdoor recreation and leisure, which made the National Park Service even more important to their states, and the growing and progressively denser urbanism in every western state, the sagebrush rebels sought fewer restrictions on the uses of public land. These latter-day states' rights activists resented federal agencies' stringent policies about grazing, timber cutting, and other forms of extractive economic endeavor. The NPS had become a particular focus of the property rights movement, with one of its gurus, an angry but articulate Ron Arnold, preposterously calling the NPS "an empire designed to eliminate all private property in the United States."[27] The fires perfectly fit an antifederal agenda. Framed as the result of bureaucratic indecision and incompetence, they lent credence to the charges of the sagebrush rebels. With a sympathetic president in the White House, one who had proven himself to be hostile to the environmental movement and its goals, western congresspeople attacking NPS policies counted on a friendly reception for their charges.

The other side of the equation for the National Park Service was the ongoing negative publicity fostered by media accounts. Many in the national media flew in and spent a day or two before writing stories that excoriated the National Park Service and its policies. Although there were knowledgeable local, regional, and national reporters, others committed the kinds of intellectual gaffes that detractors associate with the fourth estate. "People kept saying why don't you put the fires out?" Superintendent Barbee told reporters on August 23. "You don't take a 60,000- to 70,000-acre fire and just put it out. We could have had the entire United States Army in here and it would not have made any difference." The gap between the knowledge of fire management and policy and the ideas of the media that were transmitted to the public were certainly part of the NPS's problem.[28]

The NPS responded as powerfully as it could to what its staff perceived as an unjust and inaccurate set of charges. In a response to the Phase II Yellowstone Fire Report in early 1989, Superintendent Barbee offered the most direct counter to the specific charges that the NPS let prescribed fires continue to burn after the July 27, 1988, confirmation of Yellowstone's decision to reinstate suppression. Barbee insisted that the park consciously chose not to invest resources in stopping smaller fires that were in the path of larger ones if they did not threaten developed areas. Under suppression strategy, such fires fall into the "confine" category; Barbee wanted them classified as wildfires (with no response taken) rather than as prescribed natural fires. He told regional director Lorraine Mintzmeyer, "Strategically, it was decided by Area Command and agency administrators to assign all available suppression resources to those fires that posed threats to developed areas or neighboring national forest land." Even if resources had been available, Barbee assured her, "direct suppression would have made no sense and would not have been committed" to such fires. "I personally find the suggestion that Yellowstone was promoting or allowing 'prescribed natural fire' throughout late July, August, and September incredulous," he concluded. "The Yellowstone staff wants, *in the strongest possible terms*, this misperception corrected."[29]

There were supporters of the NPS, some from surprising quarters. In a powerful commentary in *Rod & Reel*, noted conservation writer Ted Williams supported NPS goals and objectives with his characteristic clear logic and incisive prose. "All the superstition about the Yellowstone fires has provided an opportunity for those who yearn to loot wild land," he told his audience. A trout advocate, he saw in the Yellowstone fires a renewing of trout habitat, a principle he extended to the rest of wild land. Yellowstone's environmental health was better as a result of the fires, Williams told his readers in a message that many of them, schooled in the conventional idea that fire was a hazard, surely found counterintuitive. He extended his argument to the NPS. "The federal government isn't perfect," he finished, "every now and then one of its agencies takes its mission seriously and proceeds with courage, intelligence, and foresight." Williams's nominee for that status in 1988 was the National Park Service.[30]

Buoyed by such support, director William Penn Mott appeared before a joint meeting of the U.S. House of Representatives' Subcommittee on National Parks and Public Lands of the Committee on Interior and Insular Affairs and the Subcommittee on Forests, Family Farms, and Energy of the Committee on Agriculture on January 31, 1989, to explain how the fires occurred and how the NPS would change its response as a result. "We must re-examine the events which led up to these fires and the fires themselves to learn all we can from them," Mott told the congressional representatives. "We can do better in similar situations in the future." Mott outlined a program of recovery that focused on fire-line rehabilitation, reconstruction of burned cabins, and other infrastructure replacement and repair for Yellowstone, Grand Teton, and Glacier. The efforts would pump $23 million into the three parks over five years, in addition to $9.1 million of emergency money for 1989. He intended to follow the recommendations of the Interagency Fire Management Policy Review Team, which had recently delivered a draft report and was compiling the public comments that derived from it. The public review of the report began in February 1989, with a final report expected soon after. Mott pointed to other changes in service policy and procedure that he said would help with the response to fire, standardize practices, and create clearer reporting and greater accountability.[31]

Outside observers felt uneasy about both Mott's remedies and the status of Yellowstone's fire management program. Some believed the park had mistakenly ignored NPS-18, which incorporated the best institutional thinking about how to make fire management happen on the ground. In the eyes of some, managers at Yellowstone seemed to have determined that their park was different. Yellowstone refused even to characterize its forests in the same language that the rest of the fire community used, preferring to invent its own idiom for describing its resources. After the 1981 season, the National Park Service convened a committee to review the park fire program; it gently urged Yellowstone to join the rest of the park system. In 1985, the regional office arranged for an experienced fire planner to spend the summer at Yellowstone in the hope that a modern document might evolve. Although the planner closely followed NPS-18, the outcome was openly flawed because the park refused to allow any written prescriptions or decision triggers that would limit the park managers' discretion and because it never submitted the revised document for public or even full agency review. Yellowstone's plan remained a 1970s-style statement of philosophy, not the manual of operations that characterized 1980s fire plans throughout the rest of the system.

After the fires of 1988, some fire scholars made trenchant critiques of NPS policy. Professor Thomas Bonnicksen, head of the Department of Recreation and Parks at Texas A&M University, was particularly harsh.

> The tragic wildfires in Yellowstone National Park have marked 1988 as the year the national park and wilderness frontier came to a close. Simply stated, shifting the responsibility or the blame to nature for the Yellowstone disaster is not an acceptable excuse. The [National] Park Service and the Forest Service are in control and they are solely responsible for their decisions. . . . The "great experiment" was the last attempt by [National] Park Service purists to retain the fantasy of a wild untamed frontier in our national parks.[32]

This characterization of the NPS as the bastion of purists defied the reality of 1988. Since the 1916 inception of the service, it had been pulled between the two different dimensions of its mandate—protection of natural resources and accommodation of the public, with accommodation the easy victor in most circumstances. Directors such as Conrad L. Wirth had been unabashed accommodators, and with Secretary of the Interior Donald P. Hodel following the precedent established by President Reagan's first secretary of the interior, James Watt, the idea that the NPS was going to let nature take its course was patently absurd. Hodel assured a national television audience on ABC's *Good Morning America* that "we are not going to let Yellowstone be damaged by" the fires. This public relations posturing belied the reality of the choices of the Reagan administration. If fire policy had escaped the efforts of the Reagan era Department of the Interior to accommodate visitors everywhere, it only was because the mantra of small government forced choices among programs.[33]

While ideological and emotional, Bonnicksen's comments reflected a particular strain of the postfire critique of the NPS. Despite the fact that his characterization of the Yellowstone situation was demonstrably false, he insisted that "wildwest [*sic*] management techniques [such] as letting fires burn unchecked" would have to change. National park lands had been altered by nearly a full century of management, he said, and were not wild, no matter how they appeared to the public. According to Bonnicksen, the "[National] Park Service in particular [was] unwilling to accept the reality that national park and wilderness areas must be managed now and forever."[34] Of course, the NPS had been managing its lands since its birth in 1916, and fire programs were always central to its efforts.

Bonnicksen clearly did not understand the constraints on the National Park Service. Quoting the Leopold Report, Bonnicksen claimed that the NPS did not recognize that park areas where suppression had been common might require "careful advance treatment" prior to the introduction of fire, although in reality the NPS had engaged in exactly that practice before every prescribed burn. In addition, at the most basic level, natural prescribed burns served almost precisely that advance treatment function for an agency that never had sufficient resources to implement a full-fledged program. Such a strategy was risky without a doubt, but it was the best available to the NPS.

Bonnicksen continued his tirade in *American Forests,* where in 1989 he published "Fire Gods and Federal Policy," essentially a distillation of his earlier arguments. Management of national parks was possible and viable, Bonnicksen insisted, but the NPS relied "instead on Mother Nature and God. In the future, managing a Park or a Wilderness will only require that rangers stand on mountaintops making incantations to the Greek God Zeus. Who needs science when you believe that the gods are managing your forests?"[35]

The subject of brutal criticism, Yellowstone superintendent Robert Barbee and a number of scientists fashioned their responses. As a natural resource specialist, Barbee had been one of the first fire managers at Yosemite more than two decades before, and he retained a powerful commitment to the principles of fire management. Excoriated during and after the Yellowstone fires, he was tagged with a nickname, "Barbee-Que Bob," and faced considerable pressure to resign. "They had a big thing over there in West Yellowstone at one of the hotels, 'Welcome to West Yellowstone and the Barbee-que,'" he remembered. In a tempered and measured response, Barbee defended NPS fire management and its goals, reiterated the value of science, and challenged Bonnicksen's judgment, values, and even his competence. In scientific terms, Barbee and his coauthors wrote, Bonnicksen lacked the clear-eyed perspective necessary to evaluate the policies and actions of the NPS.[36]

Barbee's response pleased many within the National Park Service. But because Barbee argued in the terms of science, his response could only resolve part of the problem. Despite Bonnicksen's peculiar attacks, few others believed that the NPS departed from scientific models in its management. What they disputed was the fundamental policy, the idea that some fires should be allowed to burn even if—as it seemed after Yellowstone—NPS officials were not sure they could stop any blaze once it got started.

More temperate observers offered more substantive and powerful criticisms of NPS policies and actions at Yellowstone. In Stephen Pyne's estimation, the park had unconscionably delayed in developing a meaningful fire plan. Pyne found the lack of planning to be crucial. The park still operated under the terms of its 1972 fire management plan, one of the earliest in the system. "The 1972 document was a statement of philosophy, not a working plan," Pyne insisted. Preceding NPS-18, it showed none of the influences of the new model. In the 1980s, when it seemed that every park in the system worked on a fire plan with strong operational characteristics, such as how to respond to different types of fires, Yellowstone seemed content to follow its earlier general model. Units as diverse as Pinnacles National Monument and Glacier National Park, a park that in many ways is the closest parallel to Yellowstone in the system, implemented sophisticated plans; Yellowstone did not. Barbee gracefully accepted this criticism: "The plan had been developed, but had not gone through the development process," he admitted. "I think that probably it was taking a back seat to other resource issues." Yellowstone had not been a problematic park for fire for a long time, and other pressures and concerns drew Barbee's attention away. "Fire was out there, but not stage center," Barbee concluded. "In fact, it was hardly making an appearance."[37]

The reasons were clear. Between 1972 and 1988, fire simply had not been a major problem at Yellowstone. In that fifteen-year period, 235 lightning fires burned 34,000 acres in the park. Only 15 such fires grew to more than 100 acres in size, and the largest was only 7,400 acres. The park's response had usually been swift and comprehensive. During 1979, the park experienced 29 lightning fires, 11 of which were suppressed when they threatened facilities or property. Thirteen of the other 18 lightning fires burned less than 1 acre. Even the most severe year, 1981, offered little reason to doubt the existing strategy. That year, 20,240 acres burned, an area that comprised roughly 1 percent of the 2.2 million-acre park.[38]

Barbee had faced a range of other issues between 1983 and 1988. The controversy over the reintroduction of grizzly bears continued; concerns about the removal of female bears attracted his attention; deferred maintenance issues loomed large; the NPS had recently purchased concession operations throughout the park; and as Barbee recalled, "grizzly bears, grizzly bears, grizzly bears, buffalo, buffalo, buffalo" dominated the park's horizons. In

1986, scientist Alston Chase had published *Playing God in Yellowstone*, with its acerbic critique of park natural resources management, further pulling Barbee away from issues related to fire. Yellowstone was the most visible management post in the National Park Service and one of the most complicated. Barbee and both his predecessors and successors tended to focus on the hot button issue. In the mid-1980s, that list had contained everything but fire.[39]

Two commissions evaluated the responses to the Yellowstone fires, producing two very different kinds of reports. Composed of ten people from the Departments of Interior and Agriculture, the Interagency Fire Management Policy Review Team was established on September 28, 1988, and charged with reviewing national policies and their application to fire management in national parks and wilderness and recommending responses to the problems of the 1988 season. The team quickly delivered a draft report on December 15, 1988. After a two-month period for public comments, on May 5, 1989, the team produced a final report that prompted the creation of new guidelines for NPS fire management. The objectives of the service's fire management program—the reduction of fuel loads and the use of fire as a tool to shape landscape and create a more historic ecosystem—were solid, it ruled. The report did find that the policies to reach such objectives required refinement and additional thinking and planning.

The service had to reaffirm and strengthen its prescribed natural fire policies. The report reiterated Pyne's observations: "many current fire management plans do not meet current policies. The prescriptions in them are inadequate and decision-making needs to be tightened." Further review of the plans was essential. Better dissemination of information about natural prescribed fire was a necessity, the report said, adding better interagency planning as another goal. Of the existing fire models, Alaska provided the most successful example, but even its remarkable degree of cooperation could be improved. In particular, regional planning had to be created to allow for contingency planning in extreme circumstances. A regionwide fire emergency such as occurred in Yellowstone in 1988 drew suppression crews away from home base, leaving what the review team regarded as inadequate coverage of the home areas. Internal and external NPS communication needed to improve. Many Americans still believed that the NPS let the Yellowstone fires burn to serve biological purposes, when the record showed that the NPS initiated total suppression in mid-July, a full week before woodcutters inadvertently started the North Fork fire that eventually swept through the Old Faithful complex.[40]

New fire recommendations resulted from the review team's work. On June 1, 1989, Secretary of the Interior Manuel Lujan, Jr., and Secretary of

Agriculture Clayton Yeutter directed their agencies to suppress all natural fires in national parks and wilderness areas until fire management plans that conformed to new federal standards could be developed. In addition, all fires were to be classified as either prescribed fires or wildfires, with wildfires fought by appropriate suppression methods and personnel. The responsible line officer was required to certify daily that prescribed fires were within their range and that resources to keep such fires within their prescription area during the next twenty-four hours were available. Other longer-term recommendations were included. These decisions affected not only the NPS, but also the Bureau of Land Management, the Fish and Wildlife Service, the Bureau of Indian Affairs, and the USDA Forest Service.[41] The default of the pre-1968 era had returned with a vengeance, in no small part as a result of the recommendations of a team of professionals. While it was easy to impugn the motives of the commission and to chastise the secretaries of interior and agriculture for following politics rather than science, their decision to shut down natural prescribed fires made sense in the political climate that followed Yellowstone in 1988.

A second review panel, assembled by the Greater Yellowstone Coordinating Committee, brought together scientists with backgrounds in natural disturbances. Chaired by Norman Christensen of Duke University, who had led the earlier review of fire policy at Sequoia and Kings Canyon national parks, the committee widely surveyed the many questions that surrounded the fire. "My group and my panel were given a wide mandate," Christensen recalled. "We were sort of free to go where we wanted, and we did, I think at times with a little bit of concern on the part of the Yellowstone staff people," who feared an assault on themselves, their decisions, and their policies.[42]

The science that underpinned the review process was never in question. Fire specialists such as William Romme, Dennis Knight, and Don Despain had established a historic basis for high-intensity crown fires in Yellowstone, allowing the panel to see the fires of 1988 as being part of a natural or even normal process of ecological change. In effect, existing research answered a salient question about fire in general and the fires of 1988 in particular: fire was appropriate in Yellowstone, and it did belong in the park.[43] The importance of the research and the acceptance of its data set the terms of the discussion.

While the federal review team focused on government policy, the committee of scientists reiterated a commitment to nature. "The group was always mindful of being in a situation of not wanting to create a public sense that Yellowstone as an ecosystem was in great trouble, that great ecological harm had been done by the 1988 fire," Christensen recalled. Members again asserted the importance of fire as a force in maintaining a natural landscape, memorably insisting that the "only way to eliminate wildland fires is to eliminate wildlands." Fire was an "essential component" in nature, and its removal would alter ecosystems in so dramatic a fashion as to belie the idea of natural systems, the committee declared.[44]

The commission's most compelling recommendations showed an inherent flaw in the structure of Yellowstone's fire management. Christensen believed that a "widely shared naiveté of what it is to have a natural prescribed fire program" provided a practical flaw in planning, which had contributed to the Yellowstone situation. Scientists had believed that Yellowstone National Park was large enough to comprise its own ecosystem. "If you would have asked me prior to those fires if there was any place that we might allow nature to run its course, Yellowstone National Park might be the place to do it," Christensen speculated. Fifteen years of research and management had showed that fire played an important and critical role in the Yellowstone forest, particularly among the high-elevation lodgepole pine. The experience with the prescribed fire program in the same time period, in Christensen's view, "suggested that the landscape could contain that activity." Most fires in the park during the era in which prescribed burning and prescribed natural burning had been utilized were relatively small, resuscitating an earlier myth that Yellowstone did not have big fires. Later research showed that lodgepole pine experienced fire of the magnitude of 1988 about every 300 years, with the last identifiable episode taking place in a period of high winds and extended drought in the early 1700s. The prevailing climatic conditions during most of the twentieth century seemed conducive to fires burning for short periods in small areas and then extinguishing themselves. "So, the idea that you could do this was supported by the thirteen years of data up to that point," Christensen indicated.[45]

From the comfortable vantage point after the fires, it was "easy to diagnose in hindsight," Christensen conceded, "but in retrospect there should have been a few things that caused us some alarm." The rate of burning in the Yellowstone ecosystem was very slow. Based on the thirteen years of scientific research, it would have taken millennia for the Yellowstone plateau to undergo a complete fire cycle. "We did not have on that landscape in 1987 really, really old forests," Christensen observed. "It is pretty clear that they had all experienced fire in the last hundreds of years. I suppose that might have tipped us off that maybe the experience from 1972 to 1987 was not exactly typical, or was not giving us a complete picture." But the management program for prescribed and natural prescribed fire had not been developed with this reality in mind. As a result, what Christensen called "shut-off criteria," the terms under which the NPS would begin suppression of natural fires, were not clearly defined.[46]

The lack of definition stemmed from the presumption that natural prescribed fires, those started by nature, were inherently good in a national park landscape. The only difference between natural prescribed fire and prescribed fire was supposed to be a matter of policy: when nature started the fire, one set of management precepts was invoked. The NPS did not accept ignitions that came from accidents or people in a natural prescribed fire program—a fire started from a tossed cigarette or a campfire was automatically disqualified. Lightning fires met the qualifications, but once they began, they were subject to the same rules as any other prescribed fire. All of the issues going into the development of a prescribed fire plan for a regular burn control unit would then come into play. "But in fact they did not," Christensen observed. "What in actuality happened was that very qualified people would monitor these fires and on a day-to-day basis would change in their view whether or not they were burning within prescription."[47]

This method left fire control decisions in local hands, once a goal of NPS fire management policy, but one superseded with the approval of NPS-18. Effectively, Yellowstone followed a policy from the 1970s as the rest of the service moved toward a more integrated model. From the perspective of the regional office, this was not an optimal situation. In 1985, Pyne was hired to accomplish the task of updating Yellowstone's plan. "Once I got the numbers," he recalled, "it took about five minutes to prepare a legitimate step-up plan."[48]

Pyne's presuppression work was a prelude to the regional office's real goal for the park, a new fire management plan. Pyne was expected "to nudge Yellowstone into a genuine fire plan," he recalled. "I spent ten weeks at the task and drafted a lengthy document along the lines of NPS-18" but encountered resistance from park staff. The draft plan had two critical flaws. Although it accepted "prescribed" natural fires inside park boundaries, it listed no prescriptions. "None," Pyne vehemently insisted. "The park simply would not allow anything that would limit its own discretion." Nor did Yellowstone take the plan through formal review channels. In 1988, neither public review, which was required under the National Environmental Policy Act and the Resource Conservation and Recovery Act of 1976, nor review by the NPS Branch of Fire Management had taken place. Nor did the park follow the plan. Instead, Pyne observed, Yellowstone "simply used the fact of the document to get everyone off their case."[49]

The park did not agree with or appreciate Pyne's perspective. "I ended my tour with a presentation to the park fire committee and Superintendent [Robert] Barbee," Pyne recalled. He argued for another scheme, calling for a rechartering of the entire Yellowstone fire program, on the grounds that the existing structure couldn't fight wildfires and was not suited to monitor prescribed natural fires. "The NPS was not happy. I was never invited back for another bout of planning," he said. Barbee did not recall attending the meeting but accepted the character, if not necessarily the specifics, of Pyne's assessment.[50] From Pyne's point of view, Yellowstone actively resisted the implementation of NPS fire management objectives.

From a later vantage point, some Yellowstone staff members disputed Pyne's account. According to noted Yellowstone historian Paul Schullery, "Pyne had a philosophical preference just as individual and forceful as the one held by the National Park Service managers and researchers in Yellowstone. It was just a different preference from those of park researchers."[51] The difference in opinions illustrated the diversity in possible responses and the genesis of subsequent debates about future direction.

Christensen's evaluation of the Yellowstone situation was more generous. "I sensed working with the staff, all of them from Superintendent Barbee on down, there remained a degree of defensiveness and paranoia about the program," he remembered fifteen years later. Yellowstone staff members believed that their issues were unique and that models elsewhere in the park system were not necessarily relevant to their circumstances. The combination of "all of those things led to a kind of hubris in the staff," Christensen averred, "if not certainly a defensiveness in the program."[52] To managers at Yellowstone, maintaining their discretion and prerogative was a paramount value that affected the development of park fire management policy.

Yellowstone staff saw this issue from a very different perspective. "It was more basic than that," one staffer recalled.

> Park staff simply were tired—of breathing smoke, of answering hysterical phone calls and snide media questions, of being accused of "destroying" the very place they lived and devoted their professional lives to—and by people who mostly were not there to see that, in the eyes of local beholders, . . . Yellowstone was not "gone." No one, from the superintendent down, was immune to that personal defensiveness—which doesn't make it all wrong headed.[53]

In the crosshairs of public and media expectation, under assault for circumstances largely beyond their control, park personnel could be forgiven any measure of defensiveness that manifested itself.

The degree of discretion meant that local monitoring to regulate the response to fire continued at Yellowstone without the benefit of reviewed processes and prescriptions. As a result, the monitoring program took precedence, yet the guidelines were not sufficiently substantive. "The hope had been that fire started by natural causes would simply be allowed to burn,"

Christensen remembered, but the lack of real definition of parameters made decisions about what would burn and what would be suppressed into arbitrary local ones.[54]

"Knowing what I know now, what would I have done?" Barbee rhetorically queried in 2004. "I would have probably taken action on the Fan fire; I would have snuffed out, if I could, the Clover-Mist fires. These were all lightning fires. We didn't have any [fires] that we started, and I don't think I would have done anything different on the North Fork at all." Barbee did not believe that such actions would have significantly altered the results. "Had we taken action on all these things, full suppression, there is no question in my mind the configuration would have been somewhat different, but we would have had great fires in Yellowstone. No question about it." The conditions were simply too severe, he maintained. "When you get all those variables coming out on the stage, the single digit relative humidity, and the explosively dry fuels, and then choreographed by the wind, the wind, the wind, the wind. That wind was incredible that summer," he remembered. "There is nothing that can be done."[55]

Barbee recognized that he had faced a set of forces not only beyond his control, but equally beyond those of any institution established to address fire management.

> I would have argued that let's just pull back, let's stop this nonsense of trying to draw lines around everything, let's go in and button-up values at risk, utility corridors, neighboring ranches, that sort of thing and just let [the fire] go. Because it became pretty obvious that we built lines, put in lines, worked hard, and then there were spots two to three miles ahead that burned out of control. The effort was heroic, but it was of little consequence. The joke over in West Yellowstone was "what is brown in the middle and black on both sides? A bulldozer line!"[56]

Yellowstone proved that fire management was not just a scientific process, it was equally a political exercise. Although the relationship between the two dimensions of the fire was obvious, the peculiar nature of western fire management created the illusion of their separation. The scope and scale of the 1988 fires shattered the presumption that fire was a scientific issue managed by ecological precepts. Natural prescribed fires and natural fires were ecological applications of scientific principles, generally managed by intense advance preparation and by firefighting methods that kept them under control. As long as those fires stayed within their bounds, they did not rise to the level of national policy attention. But a human-caused fire of this scale, precisely the kind against which Smokey the Bear had long warned the public, took the science experiment and placed it on a national stage, subject to new levels of review. The scale of commentary was exactly what might have been expected out of a society in which everyone competed for his or her fifteen minutes of fame. Although the NPS certainly deserved criticism at Yellowstone for the state of its fire planning in 1988, the motivations that underpinned critiques of its performance had a great deal more to do with political positioning than with the events of the summer of 1988.

The result was a wholesale change in fire management practice in the NPS and throughout the entire federal land management system. The greatest initial consequence was the effective end of prescribed natural fire and its replacement with greater emphasis on prescribed burning. Between 1983 and 1988, the National Park Service intentionally burned an annual average of 32,135 acres. In 1989, NPS-prescribed burns totaled 56,889 acres, indicative of the rising emphasis on prescribed burning that continued into the 1990s. An average of 62,843 acres was burned under prescription between 1990 and 1994. The suppression order that followed the Yellowstone fires severely constrained the amount of prescribed natural fire, which dropped from 17,944 acres per year from 1983 to 1988 to an average of 3,708 acres between 1990 and 1994. Simply put, the risk of the consequences of an escaped natural fire so outweighed any ecological or scientific advantages that might be derived from it that any reasonable park or regional office administrator eschewed the option except under circumstances that could not be disputed.[57]

At the same time, the NPS stepped up its strategic response to fire. In 1989, the acreage burned in national parks decreased by 82 percent while the number of fires in the park system diminished by less than 1 percent. A combination of more aggressive suppression and sheer good fortune accounted for the difference, but it was not clear that the change was a portent of either a more secure future or a more ecologically sound national park system. Wildfires continued in characteristic fashion, with the largest typically burning in Florida. In 1989, the year after Yellowstone, 135,494 acres burned in wildfires in the Southeast, more than 80 percent of the national park system total. Neither the prevalence of wildfires nor the 47,910 acres in prescribed burns in the Southeast excited much interest from the national press or anyone else still in an uproar about the Yellowstone fires the year before. The difference in activity, with only 10,240 acres burned in 287 wildfires and 4,993 acres in prescribed burns in the Western Region, suggested the full impact of the Yellowstone fires.[58]

By May 1990, the NPS was preparing its final report on the implementation of the changes recommended by the Interagency Fire Management

Policy Review Team. In front of the House Subcommittee on Energy and Natural Resources, NPS associate director John Morehead conceded that "a much needed tightening" of policy and procedure was necessary and "perhaps could not have been accomplished without the impetus generated by the national attention" that stemmed from Yellowstone. Yet, Morehead insisted, the possibility to overreact was great. "We must exercise caution lest we move too far toward total suppression," he reminded the subcommittee. "It is important to remember the role of fire in ecological dynamics and to ensure [that] our prescriptions maintain that natural role."[59]

Changes to prescribed burning led the implementation list. In the aftermath of the review team's recommendations, NPS special directive 89-7, issued July 12, 1989, accepted the team's report and ordered a complete review of NPS-18. The service detailed a team of NPS field and regional fire experts to the Boise Interagency Fire Center to review the fire plans for each national park.[60] By that date, the NPS had been scrutinized completely; officials at the parks and in the Washington office had begun to rethink and revise policy, and the service had compiled a five-page list of corrective actions that were already under way. Most stringent was the continuation of a new suppression policy, introduced on June 1, 1989, which was to remain in force for national parks and wilderness areas until the service judged the fire management plans for individual areas to be in compliance with the new federal recommendations.

Soon after, the Branch of Fire Management exercised the authority granted it under special directive 89-7 and offered new guidelines for policy. Finding "common management intent" with the Forest Service, the NPS developed new prescribed fire management guidelines. "A park may implement a prescribed natural fire program," the new policy stated, "only if it has an approved fire management plan" that met the criteria established for prescribed natural fire; established contingency plans for personnel and material with cooperating entities; had approved, quantified, defined prescription and monitoring procedures; detailed the availability of adequate fire management resources; and contained a process designed to outline and analyze management alternatives during a fire. This was a high standard, an effort to codify NPS fire procedures at a level never before attempted.[61]

Yellowstone was subject to the most intense scrutiny in this process. One of the first to have its fire plan assessed, Yellowstone received concentrated review. The Branch of Fire Management determined that the unreviewed draft plan of 1987 did in fact sufficiently update the 1976 plan that had been the source of much consternation in the aftermath of the 1988 fire. It required some further consideration before it could be approved and any prescribed burning in the park could resume. These technical and procedural steps were crucial, as were other reforms in practice and procedure before the plan was ready for implementation. Acting NPS fire director Douglas D. Erskine was too circumspect to point out that the lack of implementation was exactly what critics focused on in the Yellowstone debacle in 1988. Work on a new plan continued with a scoping statement, a National Park Service document designed to reveal agency intentions without committing to a plan of action, under public review in August and September 1990 and with ongoing internal critiques. The Yellowstone plan final debuted in June 1991, though questions about its practices and procedures continued.[62]

Prescribed burning at Yellowstone remained a controversial topic. Even some people in the NPS derided its prospects. Don G. Despain, one of the leading fire researchers at the park, called prescribed burning at Yellowstone "a doubtful proposition." Prescribed burning could not be "justified as ecologically necessary in most of the park," he concluded in a piece coauthored with historian Paul Schullery. "Even an aggressive program of prescribed burning launched many years ago would not have significantly reduced the acreage burned in and near Yellowstone in 1988. . . . Yellowstone's only real problem with fire is that once, every century or two, fire conditions allow more fires to burn than humans would like." Others at the park echoed such sentiments. "No plan would have altered what happened in 1988," observed Yellowstone chief of research John Varley, "and no plan will change what will happen in the future."[63]

Critics might charge that such commentary was part of an elaborate National Park Service effort to shift the culpability for the fire away from the service, but Despain, Schullery, and Varley accentuated an important and easily overlooked part of the discussion. The 1988 fires begged the question of management, a fact that NPS officials pointed out over and again to no avail. History showed that the Yellowstone environment burned at fairly regular intervals. From this perspective, it was human intolerance for such episodes that caused the uproar. Such a perspective might be scientifically accurate, but it did not mesh well with the political realities of western land resource extraction, dependence on tourism, and fire management.

The fires also highlighted a different dilemma for the National Park Service. It wanted to use science to benefit ecosystems, but politicians operated in another arena, with very different goals. Western politicians used the fires as backdrops for their various complaints about federal power and action. During the peak of the fire, Democratic presidential candidate Michael Dukakis came to Yellowstone. He arrived with more than 200 media representatives in tow, and Yellowstone had a "big thing out there at Madison Junction," a full-blown national media event. Barbee remembered, "I asked him, 'what do you want here, what are you trying to do? What do you expect from me?' He said: 'I don't want to give you a bad time.'" According to Barbee, Dukakis continued: "I will be honest with you. This is the only game in town. This is where the action is and I want some visibility and that is why I am here. I want to demonstrate my concern for the West." While Dukakis did the park no damage, his visit illustrated the difference between science and politics. "Politicians want to run things from their own perspective," observed long-time Yellowstone staff member Lee Whittlesey, "and often without regard for science, and the NPS has to listen to politicians." Most politicians neither appreciated nor understood the role of fire in an ecosystem and "that made the [NPS] task more difficult," Whittlesey concluded.[64]

Bruce Kilgore, by then chief of the Division of Natural Resources and Research for the NPS's Western Region, observed, "[E]veryone realizes [that] there are limitations to what any fire management program can accomplish when extremely dry and windy conditions occur in forests with the heavy fuel loads found in Yellowstone and similar forest types in 1988."[65] While Kilgore appreciated the impetus for policy rethinking that the situation provided, he stated what everyone close to fire management recognizes: catastrophic fire is not subject to policy constraints. Any program of management will face circumstances in which its planning, resource allocation, and procedure will be inaccurate and insufficient. Yellowstone in 1988 had been one such instance. Making policy from such an unusual circumstance was a risky proposition, but one that federal fire managers had no choice but to embrace.

By the time Morehead addressed the subcommittee nearly a year later, a new prescribed burning program had been approved for a one-year test implementation period. The revision of NPS-18 also carefully defined the prescriptions under which natural prescribed fires would be allowed to burn. The plans at the twenty-six parks with active natural prescribed fire programs were reviewed and revised to assure that they complied with the recommendations of both the review team and the commission headed by Norman Christensen, as well as the new NPS-18 guidelines. When Morehead addressed the subcommittee, four parks—Yosemite, Sequoia, Kings Canyon, and Voyageurs—had met all the requirements and were in the process of reinstating their programs. The remaining parks were expected to follow the same process.[66] Each would develop a comprehensive set of criteria to govern the conditions under which natural ignition fires would be allowed to burn, and each was expected to achieve significant progress in establishing regional and national contingency plans and procedures for curtailing prescribed fire if necessary.[67]

The acreage that burned in 1990 reflected the new strategies. Wildfires in the NPS Western Region were prevalent, with 245 fires burning across 17,732 acres during a summer when more than 225,000 acres in southern and mountain California burned during August alone. In contrast, 135 fires in the Southeast Region burned on 23,341 acres, accentuating the ways in which the issues of the post-Yellowstone era were largely confined to the West. Prescribed burns illustrated the ongoing caution. The program proceeded tentatively, and as the fire year worsened, the service brought its prescribed fire program to a halt. "At this time," acting NPS fire director Richard T. Gale told his regional and park staff on June 29, 1990, "all fire management officers should reconsider any and all prescribed burn plans, giving special attention to the limits on prescriptions that could pose control problems." The care that Gale encouraged was reflected in the program's execution. The 41 such burns in the Western Region burned 2,026 acres, a minuscule total when set against the pre-1988 efforts. Comparison with the Southeast Region provided stark relief. The 30 prescribed fires in the Southeast, largely in Big Cypress National Preserve and Everglades National Park, burned 70,396 acres. In 1992, 111 small fires in thirteen national parks comprised the prescribed natural fire total in the park system. Ninety-five percent of the acreage burned was in Sequoia, Kings Canyon, and Yosemite national parks, where experiments in prescribed burning started almost twenty-five years earlier. Only seventeen of the twenty-six parks that had a prescribed burning program before 1988 had reinstated it by 1992.[68]

In the end, one consequence of the Yellowstone fires was a less aggressive approach to prescribed natural fire in the NPS. "The revised management policies," Kilgore observed, "appear to have toned down the apparent substantial commitment to allowing natural fires to burn whenever possible."[69] Kilgore's observation reflected the dismay current among specialists who administered fire at the grassroots level who were forced to abide by the post-1988 rules. For many in the fire management structure, it was hard to see the change in goals as an advancement of NPS principles. The default to suppression flew in the face of twenty years of experience.

The greatest challenge to the renewed ethic of suppression took place in Alaska. After a number of unusual years in which the Alaska parks experienced little or no fire, blazes returned in 1990. That year, seven large fires in Alaska burned more than 108,722 acres, hardly a record in the nation's northernmost state, but a harbinger of management issues that challenged the NPS's vision of what it could do with fire in the far north and elsewhere. When the total burned in Alaska in 1991 reached 86,651 acres, the NPS was forced to address the substantial differences between the forty-ninth state and the situation in the western part of the lower forty-eight.[70]

The Alaskan response to the review team's recommendations had been outrage. Both state and federal land management agencies had recognized the efficacy of natural prescribed fire and were committed to its continued use. NPS officials could say little after the public outcry around the Yellowstone fires, but state officials loudly highlighted the differences between the lower forty-eight and the north. "The state of Alaska takes objection to the review team's recommendation that it is 'unprofessional and impractical for fires to be allowed to burn free of prescriptions or appropriate suppression action,'" the official response of the Division of Forestry of the Alaska Department of Natural Resource intoned.

> We must express that there are regional differences across the nation in natural resource management goals and natural fire regimes. Alaska is a *fire dependent* ecosystem. There are vast areas of Alaska where fire does not pose a threat to people, property, or valued resources. In these places, fire is viewed as a *natural event.*

Alaska state forester M. R. "Bob" Dick, Jr., asked NPS regional director Boyd Evison to "carefully review any change in national fire policy that would compromise the ability of Alaska-based federal agencies to continue with current fire management plan arrangements. Put bluntly," Dick concluded, "if it ain't broke, don't fix it."[71]

Dick's trenchant observations illustrated that the NPS grappled with wildland fire in two and even possibly three dramatically different situations. Alaska shared many parallels with the Everglades, Big Cypress, and the Southeast Region, themselves significantly different from the western fire parks, but in the end, the forty-ninth state was substantially dissimilar from even the closest examples in the lower forty-eight. Fire in Alaska was perceived by land managers as an irresistible force that could overwhelm the resources to battle it at any time. Alaskan fire, in the view of the conglomeration of state officials and federal agency managers who administered the vast territory, was natural fire at its most distinct, a clearly defined natural force that humans could not genuinely conquer or even trifle with, except at great peril and greater cost.

The result was a strategy simultaneously more open and more defensive than the one the NPS applied at Yellowstone. The openness stemmed from the acceptance of natural fire's value as a primary ecological tool and from the tacit admission that there was little that could be done about most Alaskan fires anyway: they would burn and that was an ecological and social good. The defensiveness resulted from the peculiar governmental circumstances in Alaska—the multiple layers of federal, state, native, and regional agencies and their varying missions. In Alaska, federal agencies could not genuinely expect to implement a comprehensive suppression policy, and no one associated with Alaskan fire believed that they could.[72]

In a 1991 response to such concerns, the NPS dispatched a team to discuss long-range fire management objectives for Alaska. NPS fire director Douglas Erskine and fire management specialists Richard T. Gale and Rod Norum reviewed a proposal from regional fire management officer Steve Holder and regional prescribed fire specialist Brad Cella to establish an Alaska Wildfire Coordination Group. The reviewers discovered "a renewed and vigorous interest in the use of management ignited prescribed fires" among land management agencies in Alaska, and they advocated enhanced planning, programming, and interagency cooperation. Their work affirmed a principle that resonated in Alaska and that the NPS embraced in the far north. Suppression decisions "will remain with the local land manager," Erskine informed deputy commissioner Ron Somerville of the Alaska Department of Fish and Game.

> Because the DOI [Department of the Interior] agencies manage their lands on multiple use principles, we have received favorable interpretations that wildfire surveillance is an appropriate response if it is allowed in an approved plan and determined to be the most cost effective action. . . . The DOI agencies have elected to use their Alaska fire management personnel to implement this fire policy for Alaska rather than accept a very narrow national definition.

By articulating the differences between the Alaskan experience and the rest of the nation, the NPS built stronger ties with state and local agencies and interests.[73]

After the early 1990s, fires in Alaska diminished enough that they were no longer in the forefront of national policy discussions. After 40,035 acres burned in twenty-six fires in Alaska in 1993, the burned acreage diminished to a low of 21 acres in 1998. Only in 1999, when fourteen fires burned across 164,397 acres did Alaska again attract policy attention.[74]

During 1990, as large fires burned in Alaska, the NPS faced the first real challenge to its reconstituted fire management in the lower forty-eight states at Yosemite. 1990 became brutal for fire in California, affecting the service's ability to battle fires and all but eliminating any efforts to reintroduce prescribed fire in the western parks. As Yosemite prepared for the centennial of its establishment, the park was closed for the first time in its history. Lightning storms ignited fires in the park between August 7 and 10. Suppression efforts ensued, but failed to halt the spread. By August 10, more than 12,000 acres were aflame. Yosemite Valley and El Portal were evacuated that day, and the Merced Grove of big trees was threatened. Although the fires continued, they were brought under control later in the month, and the episode seemed likely to pass without renewing the controversy that had dogged Yellowstone.[75]

Later observers noted that the situation at Yosemite in 1990 roughly paralleled Yellowstone in 1988. In both fires, severe drought contributed to the accelerating danger. In both instances, high temperatures and low humidity combined with thunderstorms to intensify the fire risks. But the two parks are different both in their ecology and their vision of the role of fire. All of the damaging fires at Yosemite occurred outside of prescribed natural fire zones and were automatically subjected to suppression. Yosemite had long recognized that fuel conditions in the mixed-conifer forest and chaparral zones were not within the natural range of variability and that lightning fires would not be ecologically beneficial. NPS responders were initially overmatched, but the arrival of Type I and Type II incident management teams gave pivotal assistance. The relatively small size of the 1990 Yosemite fire—between 12,000 and 15,000 acres—and the combination of skilled personnel and changing weather helped to bring the fires under control.[76]

Widespread media attention added to the service's caution at Yosemite. NBC, CBS, ABC, and CNN were all present; "Good Morning America," the "Today" show, *Newsweek*, and *Time* all covered the fire; and countless local, regional, and national newspapers tracked the NPS's response.[77] By 1990, fire management had become national news, and the National Park Service again found itself at the center of a public debate about how to respond. Yellowstone and Yosemite both experienced significant fires at a time when policies and public perception did not coincide. This brought the national parks further scrutiny. Even more, changing patterns of living and a callous ignorance of fire patterns had brought growing populations into potentially threatening environments, sometimes near or adjacent to national park areas. It only remained a matter of time until hillside suburbs around arid western communities faced the brunt of fires of their own.

In the early 1990s, the West turned mildly wet for a few years. The result was a clear decline in acreage burned by wildfires. The Intermountain Region of the NPS recorded fires on 30,750 acres in 1990, 7,776 acres the next year, 3,744 acres in 1992, and 14,400 acres in 1993. Fire damage in the Pacific West Region decreased from 20,616 acres burned in 1990 to 6,342 acres in 1991, 11,468 acres in 1992, and 8,788 acres in 1993.[78] As a result, fire concerns shifted from national park fires to those on the lands of other agencies.

The West's dramatic and brutal fire year of 1994 drew the issue of fire management even further from the National Park Service. In the first bad year in the region in five years, considerable NPS lands burned—52,502 acres in the Intermountain Region and another 20,565 acres in the Pacific West—but the acreage became minuscule in comparison to the human cost. The real story of 1994 became the horrific deaths of thirty-four firefighters in the line of duty, and $965 million was spent on suppression as fires burned on 3.5 million acres. In one tragic afternoon on July 6, 1994, in the South Canyon fire on Bureau of Land Management land outside of Glenwood Springs, Colorado, twelve firefighters and two helicopter crew members, trapped as a fire swept upslope, burned to death. Pyne opined that "the firefighters lost at the South Canyon fire were, for the fire community, the equivalent of the Army Rangers killed at Mogadishu." Those tragic human losses inexorably altered both policy and procedure.[79]

In the aftermath of the tragic summer of 1994, the National Park Service again reassessed its fire management strategies and goals. A study team of Stephen J. Botti, G. Thomas Zimmerman, Howard T. "Tom" Nichols, and Jan van Wagtendonk, all respected fire researchers or managers, analyzed NPS fire problems. They advocated increasing the amount of park acreage that functioned as a natural ecosystem; reducing the risk of severe wildfire in developed areas in national parks and along boundaries by clearing and prescribed burning; enhancing efforts to provide information about the role of fire in parks to the public and to decision makers; increasing interagency planning; and increasing the capability to analyze data and integrate fire management into the general planning and management of the park system.[80] The recommendations reflected the set of goals the NPS had developed prior to Yellowstone, with a strong dose of the programmatic changes that followed the 1988 fires. Yet what made the report significant was its willingness to assert the value of natural fire in the aftermath of the summer of 1994. The NPS continued to embrace the role of natural fire even as such a stance became politically more difficult to maintain.

During the summer of 1994, two initially small fires at Glacier National Park attracted regional attention because the NPS was willing to let them burn. "In a year when firefighters are scrambling throughout the West, Glacier National Park managers are carefully tending a small 6-week-old fire that could potentially burn a whopping 43,000 acres, maybe twice that much," wrote Don Schwennesen of the *Missoulan* in an overstatement of the potential impact of the fire. Such media attention could easily shrivel a fire manager's desire to support prescribed burning. Even Glacier superintendent Dave Mihalic seemed to vacillate, noting the enormous difference between a policy and its implementation. "While the public may support prescribed natural fire in theory, such discussions typically occur outside the fire season," Mihalic told the *Billings Gazette*. Actual fire made such support tenuous at best. Nor was the park helped by a flippant comment from Flathead National Forest spokesman J. D. Coleman, who told reporters that the Forest Service "was not screwing around with prescribed fires right now." The internal and external struggles over the fire continued into the middle of August, with the NPS closely monitoring the fire even as local and regional reporters queried local residents about the policy. Rain on August 18 slowed the fire, and snow and rain in early October finally removed the threat.[81]

In the end, one of the two fires, the Howling fire, proved to be a triumph for prescribed natural fire. Although the fire was routinely termed "controversial" by regional media, the pressure on the National Park Service was not sufficient to compel suppression. The service stuck to its principles despite the discomfort it caused local managers. "If we would have put the fires out just because of the [fire] activity around us and political developments," observed Glacier National Park fire management officer Fred Vanhorn, "we, in effect, [would be] saying that we were not going to allow significant prescribed natural fires to occur in Glacier." NPS staff felt that the Howling fire provided an important lesson that could be used as a model elsewhere in the park system. By holding firm to established policy even under political pressure, Glacier National Park proved that prescribed fire could take place, even when fire was a severe problem elsewhere in the region or in the park system.[82] This was a step toward the more comprehensive vision of the role of prescribed natural fire held by most fire scientists and ecologists inside and outside of the national park system.

The National Park Service continued to develop and support its prescribed natural fire program. In an important workshop in San Francisco early in 1995, the NPS reaffirmed its commitment to the concept. The period following 1988 had been marked by an aversion to the risk of an escaped prescribed natural fire. With the minimization of risk as a goal, the burn program could be scored a success. But if the goal were to implement prescribed natural fire programs that were ecologically significant, "pretty disappointing" was a better description of the situation. "We need to find a way for prescribed fire and wildfire programs to coexist during the normal fire season," the meeting summary recorded. Such a strategy would require expanding prescribed natural fire while maintaining an acceptable level of risk.[83] This was as difficult as the NPS agency mission: to preserve for the future while accommodating the present.

On the ground, new innovations revealed new approaches to fire. In 1995, the NPS introduced its Prescribed Fire Support Module (PFSM) program. The PFSM program provided mobile tactical support nationally for parks with prescribed fire programs. Because they were specifically unavailable for wildfire response, the NPS teams, initially consisting of four groups of five members each, guaranteed experienced professional attention for prescribed fire. Subsequently, similar teams designed to manage prescribed natural fire were introduced. When the NPS lifted the budgetary ceiling on prescribed natural fire, it effectively removed the rationale for converting fires from the prescribed natural fire category into the wildfire category as a way to access resources. This development further contributed to the growing role for natural prescribed fire in the NPS.[84]

The idea that the National Park Service would risk potential political pressure in the name of a clear ecological and resource management goal spoke volumes about the commitment of the service to the goals of the previous twenty-five years. Despite the enormous negative publicity associated with the Yellowstone fires and the intense scrutiny and micromanaging that the NPS experienced in its aftermath, the service had a vision of an appropriate strategy and was willing—within reason—to take risks to implement it. At a time when morale was low throughout the National Park Service, due in large part to director Roger Kennedy's 1995 reorganization, which redistributed power away from skilled specialists, this firm stand on principle proved inspirational to many.

In 1995, a revised national fire policy was completed. The NPS played an important role in shaping the document. A direct result of the 1994 fire season and the South Canyon tragedy, the new document articulated nine management principles that sounded much like the NPS's goals. Under this document, public and firefighter safety remained the top priority. Wildland fire was seen as an essential ecological process and an agent of natural change that had to be incorporated into planning. Fire management plans were required for every federally administered area with vegetation that could burn, and fire planning had to be designed to support land and resource management planning. Risk management became a foundation for fire management. Fire management programs had to be economically feasible and had to be based on the best available science. The plans had to incorporate environmental quality and public health considerations. Cooperation with other governmental and nongovernmental entities was essential, and the standardization of policy among federal agencies was to be an ongoing objective. The plan emphasized indirect attacks on fire, the sort of approach that had characterized the response to the Yellowstone fires in 1988, as a safer strategy than the direct confrontation of firefighting lore. A full range of responses was permitted, allowing a measure of flexibility that had never before existed across the full spectrum of federal agencies.

Any approach, from basic monitoring to full-scale suppression, could be implemented under the right circumstances, a substantial shift in the way federal agencies approached fire. In effect, the federal system changed from looking at the origin of a fire to looking at its circumstances as the basis for decision making.[85]

The new policy led to greater coordination and cooperation among federal land management agencies. An implementation plan in 1996 translated the vision into a series of programs, dividing the initial recommendations into geographic concerns and long-term commitments. The difference suggested the enormity of the task that confronted federal fire managers and the need for clear signposts—not only to gauge progress but also to remind everyone of the direction in which fire management planning was going. Ongoing policy reviews and innovation led to further planning and new goals. A new resource allocation strategy in 1998 sought to develop an interagency process to distribute fire management resources while efforts continued to move fire management policy toward implementation throughout the national park system.[86]

Between 1995 and 1999, federal agencies more than doubled the acreage treated with prescribed burning, which reached 2.2 million acres as the new century began. The NPS was a small portion of that new emphasis, burning 59,495 acres in 1995, 42,511 acres in 1996, 69,481 acres in 1997, and 82,413 acres in 1998, finally topping the 100,000-acre total in 1999 with a combined burned acreage of 132,665. With the exception of 1999, the second half of the 1990s showed no significant difference from NPS prescribed burning during the first half of the decade. The shift in national emphasis had little impact on NPS practice, leading to questions about whether the bold promises of the mid-1990s amounted to significant changes in practice.[87]

The National Park Service continued to develop strategies for addressing fire. In 1996, NPS fire director Doug Erskine pointed to a significant improvement in the tools available "for expanding the use of fire in national parks." Prescribed fire modules had been thoroughly tested and proven successful, and their use had been expanded. In 1996, the NPS established four prescribed natural fire management teams, with a planned increase to six in 1997. Dedicated fire specialists were located in the Midwest, Intermountain, and Pacific West regions, and the Southeast and Northeast field areas shared another team. A significant change in funding allowed further development of the NPS response to prescribed natural fires. Congress permitted the NPS to fund the operational aspects of prescribed natural fire from the suppression fund. Combined with the endorsement of the Federal Wildfire Management and Program Review (FWMPR), designed to cross agency boundaries and to be based on the best available science, and of the secretaries of the interior and agriculture, the NPS could make the case that its approach to fire had shaped federal policies.[88]

Throughout the remainder of the 1990s, fire policy remained an important component of federal land management that largely embraced the principles the NPS had developed since the 1970s. In a perplexing turn of events, national park lands were not the focus of the program, something for which the National Park Service could be grateful. A truly national system had developed, one focused on firefighter safety, land restoration, and federal lands other than the park system.

EIGHT The Hazard of New Fortunes

Outlet, Cerro Grande, and the Twenty-First Century

The National Park Service could be forgiven for thinking that its efforts to institute fire management during the last twenty-five years of the twentieth century were cursed. After Yellowstone in 1988, NPS efforts to mitigate fire and to plan for its management throughout the national park system met with great success during the next decade. The service evaluated its response to fire, designed new mechanisms to bring practice and ideology into a coherent relationship, and invested resources in internal responses and in interagency planning, resource acquisition, and deployment. By the late 1990s, fire managers felt that they could view their very complex field with a little more ease. Yellowstone initially threatened the very stability of the National Park Service, for the fires of 1988 called into question not just NPS practice, but the very rationale for managing park areas. The development of a national fire planning and management structure—the new standards that the NPS successfully implemented and the remarkable biological renewal of Yellowstone—combined to give the service's fire management greater credibility with the public than it had ever before enjoyed.

The urban fires of the 1990s, especially the Oakland–Berkeley Hills fire in 1991, actually improved the service's status. Even though these fires were remote from national park lands, they heated up discussions about managing fire even as they highlighted the successes of the National Park Service. Voices commenting on those fires joined public intellectual Mike Davis, whose postapocalyptic *Ecology of Fear: Los Angeles and the Imagination of Disaster* reached number one on the *New York Times* bestseller list as it sparked a controversy over whether communities that built in hazardous fire regions merited the response of public services. One chapter, "The Case for Letting Malibu Burn," spurred particular animosity even as it directed much of the rancor about fire away from the National Park Service. Davis argued that communities that approve home construction in clearly hazardous locations deserve to face fire without the deployment of external resources. In so doing, Davis shifted the debate over fire management from federal agencies to cities and counties.[1]

Davis's work reflected the increasing proliferation of what the historian Lincoln Bramwell called "wilderburbs"—communities that emerged at the point of collision among rural land, wild land, and urban expansion to take advantage of the natural amenities around them—and ignited a debate about the selection of locations for new communities in the posturban West. Even as federal agencies removed fallen trees and underbrush on more than 2.2 million acres of land in 1999 alone to limit the chances of fire, more than 200 million acres historically prone to frequent fire carried the heavy underbrush associated with suppression. The result was dangerous and left not only federal land managers but also officials at the state, county, and even local levels in a difficult situation.[2] Urban and suburban sprawl had become endemic nationwide; in the West, it encroached on national park areas and added a range of new problems for their managers. Much of the region lacked adequate mechanisms to regulate suburban planning. As a result, communities grew nearly everywhere, adding the threat of accidental or intentional fires from nearby populations to the problems of park managers, as well as the possibility that such communities, located without much more than a nod to safety from wildfire, might very well serve as conduits for the inevitable fires in a region's drier sections. Even as the National Park Service grappled with urban parks such as Golden Gate National Recreation Area, urbanization and its attendant sprawl encroached on previously remote or distant national park areas in the West and throughout the nation.

By the mid-1990s, the National Park Service had achieved the respect of the fire community; the NPS model had become the currency of federal fire policy. The importance of the NPS philosophy became clear as other federal agencies recruited NPS fire personnel for their agencies, a reversal of the sixty-year trend of personnel movement from the Forest Service to nearly every other federal land management agency.

At the same time, a long series of drought years in the West that began in the mid-1990s contributed to a critical change in regional fire patterns. The growing wilderburbs problem shifted the focus back toward conventional models of fire suppression. By 2000, the Forest Service had regained an important measure of its earlier position. Many of the major fires were on its lands, and its holdings included many of the places where wildland and urban growth coexisted so uneasily.

As spring turned to summer in 2000, a pair of nearly simultaneous prescribed fires on national park lands escaped their bounds and illustrated the gravity of the changes occurring in fire management and the fundamentally tenuous nature of all existing strategies to manage, combat, or regulate fire. The emphasis on prescribed burning that characterized the period after the new national fire plan in 1995 yielded tremendous benefits for the NPS, but contained parallel risks. The acceleration of prescribed burning programs treated considerable acreage, but as always, a great deal of land experienced no such management. The reasons varied; in some cases, prescribed burning was deemed too dangerous because of the proximity of communities, facilities, and other development. In far more instances, the resources were insufficient, the time to undertake such a program too short, or the conditions deemed unsuitable. The NPS treated as much land as it could and planned for more, all the while recognizing the inherent risk in any program that allowed fire in any way. With the ongoing drought and some simple bad fortune, risk came to the forefront in May 2000.

At Grand Canyon National Park, the Outlet fire near North Rim Village illustrated one of the fundamental problems with introducing fire. By 2000, the Grand Canyon had engaged in a prescribed burning program for almost two decades. A fire management plan, approved in 1992, had been revised each subsequent year. These reviews looked at program successes and identified areas of concern. A 1997 NPS review team identified a long-standing problem in the North Rim forests: an accumulation of litter, such as fallen trees, branches, leaves, and needles. Combined with an invasion of spruce and fir, which provided fuel ladders that could lead to enormously destructive crown fires, this set up a potentially dangerous situation. The problem first had been recognized on the North Rim in 1981 by regional plant/fire ecologist Kathleen Davis and was reiterated during the 1990s. The 1997 review team recommended the development of a landscape-level prescribed burning program for the North Rim forests.[3]

The team of experienced fire managers who undertook the 1997 review—Steve Botti, Jim Douglas, Steve Tryson, and Wally Josephson—recognized the North Rim as an example of the problems of fuel load accumulation that vexed so much western wildland and that remained the subject of powerful debates in the professional and scholarly fire communities. The team recognized that the existing prescribed burning program on the North Rim had only achieved some of its stated goals. The suppression of two prescribed burns that escaped, the Mathes and the Northwest III fires, along with concerns about smoke, which marred visitors' experience, led to what the review team described as a "conservative approach" to the reintroduction of fire as a management tool at the park. Yet, the team found Grand Canyon staff willing to be more aggressive in their response than they had been in the past.[4]

The Grand Canyon had been thoroughly studied by a range of scientists, and different schools of thought offered their own remedies for the North Rim problems. Headed by Professor W. Wallace Covington of Northern Arizona University, an experienced fire scholar who focused his research on the Grand Canyon, some researchers believed that because of the heavily loaded ponderosa pine–mixed conifer forest, the North Rim required mechanical thinning of the biomass before the introduction of prescribed fire. In deference to this research, the review team suggested that "testing the truth of this hypothesis should be a central component of the park's fire management program over the next five to ten years." The advantages were obvious: reduced risk to land and people and an opportunity for the prescribed fire regime to mirror natural fire rather than replace it when it did not occur. This fire management approach provided opportunities to assist local residents with timber sales and other economic advantages of such cutting and helped to avoid the social and political fallout that typically accompanied smoke emissions. The disadvantages of removing as much as 90 percent of the North Rim's forests included the likelihood that visitors would balk at what would certainly seem to be a denuded landscape; the problems that logging might cause for wilderness designation and wilderness study areas; the threat of severe wildfire before the thinning could take place and the attendant problem of burning the slash that would remain; and the many stumps that would be visible to a traveling public that held a decidedly different view of what a national park should look like. The review also noted cost concerns. Even with timber sales and slash logging, the expense might be prohibitive.[5]

After weighing the conditions of the situation and possible solutions, the team made clear recommendations for a more aggressive prescribed burn policy. Team members believed that using "fire as a management tool offers the best hope for preventing catastrophic wildfire and restoring the natural ecosystem in the long run." Resources were available for hazardous fuel reduction, and the Grand Canyon staff had begun to use them to carry out large-scale burns when weather and other conditions permitted. The team also recommended using nearly every "tool in the toolbox"—mechanical thinning, planned ignitions in key areas, and the ideas of Covington and other scholars about the impact of fire on native plants. The park adopted the report's ideas in a 1998 revision of its fire plan.[6]

In accordance with that plan, on April 25, 2000, Grand Canyon staff ignited the Outlet prescribed burn in a region of mixed conifer and piñon-juniper complex around 8,400 feet and just west of the developed area on the North Rim. The goals of the prescribed burn were to perpetuate natural processes and to reduce hazardous fuels. On April 27, another fire, on the Tiyo Subunit, suffered an incomplete ignition that resulted in a "dirty burn," fire management lingo for a fire that did not go as planned, but it remained inside its prescription area. On May 8, firing began on the east side of the Widforss Subunit of the Outlet fire. The burn proceeded in accordance with expectations until the next day. A wind came up on May 9, and an undetected spot fire on the Widforss Subunit grew until it exceeded the parameters of its prescription. This threatening situation drew scrutiny from park officials, and later that day, when weather experts predicted strong winds for the next day, Grand Canyon superintendent Robert L. Arnberger declared both the Tiyo and Outlet as wildfires. He alerted the Type II Northern Arizona Incident Management Team, headed by incident commander Larry Anderson, and asked it to be ready to take over fire suppression efforts.[7]

On May 10, the expected major wind event took place, changing the Grand Canyon's response to the Outlet fire. Gusts reached sixty miles per hour, and by 2:30 p.m., the park was evacuating the North Rim of the Grand Canyon. Snags and fallen green trees blocked roads, and fire crews were forced to bed down in the open to avoid hazards such as falling trees. By May 11, the fire was estimated to cover between 1,500 and 2,000 acres, but much of the burning land was close to the developed areas at North Rim Village. This had potentially severe consequences. At the Nankoweap trailhead, visitors were stranded beyond the fire and hemmed in by downed trees, adding a search-and-rescue dimension to the obligations of the incident management team when it took charge of the fire.[8]

The single largest problem that the incident team faced was a shortage of suppression resources at the Grand Canyon. Firefighting personnel were in short supply, with only two interagency hotshot crews and two Type II crews available. "A couple of Type 6 engines, 3 prescribed fire modules, and a Type 2 helicopter that couldn't fly in the high winds" were all the resources available, Al Hendricks of the Northern Arizona Incident Team wrote. The fire had grown much larger by the time the Type II team took charge, with aerial observation indicating that it had reached 7,000 acres. Plans to call in a Type I team gained momentum, with input from the Washington office of the National Park Service. The Type II team established a base camp in the Kaibab National Forest, just outside the park boundary. The winds died down on May 12 and 13, and the arrival of the Northern Rockies Type I Incident Team headed by Steve Fryes helped to provide the work power to initiate suppression. "We can fight this fire aggressively," Fryes told the press, "but we first do it safely and with sensitivity to the area's natural and cultural resources." By May 13, when the Type I team took control of the fire, a full complement of suppression resources was available.[9]

During the next week, suppression remained the sole mode of response. Stronger winds and low humidity aided the fire's growth on May 14, but 800 firefighters continued to dig hand lines, providing 20 percent containment. High winds on the following day contributed to erratic fire behavior. The blaze did not cross any established control lines that day, and the park reported containment at 43 percent. On May 16, the fire grew to 13,350 acres, even as the total force fighting it reached 914. Favorable weather conditions on May 17–18 helped crews start to gain control of the fire, and with sixteen miles of hand lines completed and as many as six left to dig, the fire was reported to be 56 percent contained. Superintendent Arnberger expressed his support and gratitude for the work of the fire crews. On May 19, he declared that the North Rim would reopen to visitors on Monday, May 22. By Sunday night, incident commander Steve Fryes could report total containment of the Outlet fire.[10]

After the incident, the assessment of the fire showed little culpability on the part of the Grand Canyon. An investigative team, led by cochairs William F. Paleck, superintendent of North Cascades National Park, and forest supervisor Rodd Richardson of the Bitterroot National Forest, generally praised NPS's preparation and handling of the fire. The "overall competence, professionalism, and accomplishments" of the Grand Canyon prescribed fire program were unassailable in the investigation team's assessment. The problems it identified—that fire leadership was, in the words of the report, "spread too thin for too long" and that the plan did not contain enough contingency triggers in case of escalation—were minor. "The prescribed burn program at Grand Canyon National Park is fundamentally sound," the report concluded. "Continuation and even expansion of current program levels is absolutely necessary to safeguard the park from the effects of nearly 100 years of fire exclusion."[11]

The Outlet fire was dramatic. Its small size and disproportionate impact served to illustrate the growing dilemma of the twenty-first-century American West, where formerly wide-open spaces are increasingly dotted with people. The effects of ecological restoration have to be deftly balanced with the needs of adjacent communities and those of travelers, concession holders, and others. Even the most adroit calculation can spiral out of control, as the Outlet fire on the North Rim of the Grand Canyon demonstrated.

At the same time, the review highlighted the ways in which the problems with fire are inherent and random. The Outlet fire could have happened anywhere at any time; its jump out of bounds resulted not from bad planning or decision making, the review committee concluded, but from changing natural conditions. Unlike earlier fires, where critics pointed to flawed policy or mistakes in implementation, at Outlet the NPS made no significant errors in planning or operations. Fire management includes risk; the assessment of that risk is more a political and cultural question than an ecological one.

At about the same time, a second prescribed fire escaped its control lines. The Los Alamos area had been the scene of a number of fires that were frightening more because of their potential impact than their actual size. La Mesa in 1977 had been the first significant fire since the Manhattan Project built the atomic bomb in the 1940s. Later fires in 1996 and 1998—the Dome and Oso fires—illustrated the dangerous result of the long-standing practice of excluding fire in the vicinity: a heavily fuel-laden region with an urban area at its core. Studies by ecologist Craig D. Allen and dendrochronologist Tom Swetnam suggested that the last thorough burn in the vicinity of Los Alamos took place in 1881. This century-long suppression compared unfavorably to an analysis of the eighteenth and nineteenth centuries, when significant fires occurred about every six months. What made the situation particularly dangerous was the proximity of the town of Los Alamos and the Los Alamos National Laboratory, where significant experimentation with weaponry still took place and where the U.S. government had stored radioactive and explosive materials. Although everyone knew that Los Alamos housed weapons research and contained an array of dangerous compounds and chemicals, national security constraints restricted the details of that information. Firefighters had little idea of what they might encounter.[12]

When NPS officials from Bandelier National Monument authorized a prescribed fire for the Upper Frijoles Creek drainage on April 19, 2000, they could not have anticipated that the fire would be the catalyst for another reevaluation of fire in the national park system. Park officials selected the Upper Frijoles drainage units 1 and 5 for the burn. An earlier effort to burn unit 1 in 1993 had not achieved desired results, but in accordance with the park's fire management plan, efforts to burn these areas continued. The 1993 fire had only minimally diminished the fuel load of 34.4 tons per acre before the burn, to 29 tons per acre, a 16 percent reduction. The primary purpose of the 2000 burn was to further reduce hazardous fuels in the units. A three-part approach was approved. The park needed dry conditions to achieve its goals everywhere but in the high-elevation grasslands. The planned first phase was to burn the upper area that included the grasslands; the second phase was to burn the timbered areas and the drier, south-facing slopes throughout the area. Managers planned to delay the third phase until the wettest areas of the units were dry enough to burn.[13]

Bandelier's prescribed fire initially seemed to be an ordinary event. After a May 4 amendment to the plan, which excluded private property on the

Valle Grande from the project, the burn boss, Mike Powell, made the appropriate notifications and conducted the required briefings. At 7:20 p.m. on May 4, 2000, crews ignited a test fire atop Cerro Grande in the Jemez Mountains in the westernmost part of the park. By 8:00 p.m., the test fire was completed and officials deemed it to be within prescribed parameters. By 10:00 p.m., crews completed the ignition process on the northeast edge of the fire area. At 11:15 p.m., ignition of the northwest area began.[14]

By the early morning of May 5, the Cerro Grande prescribed fire had begun to spread beyond its intended boundaries. At 1:00 a.m., crews reached the upper saddle of the mountain and spent ninety minutes bringing the fire back into the saddle. Ninety minutes later, the fire seemed under control, and the burn boss and crews stopped to rest. At 3:00 a.m., the burn boss asked for help from a Type III helicopter and a twenty-person hand crew. Although it was not customary to order helicopters for prescribed burns, Bandelier's fire management officer and the zone dispatcher agreed that the resources, ordered for a wildfire burning in the Santa Fe National Forest, would be diverted to the prescribed burn on national park land.[15] Although this was a little unusual, officials believed the burn still seemed to be manageable with little more than typical resources.

At 10:00 a.m., conditions seemed more threatening. Wind changes initially created some spotting within the designated area, and slopover on the upper east part of the fire line caused some concerns. The crew on the northeast side reported difficulty in containing the fire within the designated boundaries. Managers requested water drops and extra firefighters. At 10:30 a.m., a helicopter dropped two people off on the northeast side of the fire and departed to pick up the bucket and begin water drops. At 11:00 a.m., the Type I hand crew arrived; five of its members headed to the west line, while the remaining thirteen went to the troubled northeast side. An air tanker, a cargo airplane modified to deliver a water or chemical retardant payload, was requested for the slopover, and it arrived at 12:55 p.m. Five minutes later, Powell converted the prescribed fire to a wildland fire, changing his status from burn boss to incident commander.[16]

During the subsequent thirty-six hours, the regional fire apparatus responded. Paul Gleason, an experienced NPS fire manager, took over as incident commander, and he briefed park management on the renamed Cerro Grande fire. He, Bandelier's fire management officer, and the chief of resource management designed a Wildland Fire Situation Analysis (WFSA). Gleason made several critical tactical decisions. He decided that crossing the face of the mountain was too dangerous. "I gave to the park superintendent, as my preferred alternative, to go indirect, down to Route 4," he recalled in a panel discussion about the fire. Crews improved the existing fire lines and used drip torches to widen the burned areas aimed at slowing the fire's progress.[17]

At first, the strategy seemed to work. A National Weather Service spot forecast at 11:55 p.m. on May 5 had called for a fire weather watch the following day, but with the resources available, Bandelier managers believed that preparations seemed adequate to the rapidly changing situation. Spot fires outside the designated boundary were contained on May 6, and blacklining continued. A meeting among the park's staff and representatives of the Los Alamos National Laboratory, the U.S. Forest Service, and Los Alamos County addressed suppression strategy and tactics in the WFSA.[18]

As late as 11:00 a.m. on May 7, most spot fires appeared to be contained, and the day began with optimism. At 2:30 a.m., the fire had been tied to its anchor point at Route 4. Bandelier's next objective was to bring the blacklining fire across from east to west, but the wind blew downslope, exactly the wrong direction for such a goal. Even though Frijoles Canyon, choked with fuel according to one description, sat across the road, fire managers felt compelled to wait. The situation still seemed manageable. Just as officials felt that they had contained the fire, west winds dramatically increased, and the fire spread into the adjacent Santa Fe National Forest. By noon, the fire had spread south of Route 4 into the Upper Frijoles Canyon drainage, burning with an intensity that made it impossible for the crews to attack it. The blaze turned into a crown fire, sending embers flying ahead that created spotting and crowning east of the prescribed fire zone. As this fire broke containment, a Type I incident management team was ordered. At 12:40 p.m., Gleason, the incident commander, ordered the evacuation of Graduation Flats and American Springs; shortly after, all agencies in the area closed roads on their lands and evacuation procedures began.[19]

By 3:00 p.m., the situation had turned even more dangerous. East of the burn, watchers reported spot fires with the potential to threaten the town of Los Alamos and all of its installations. Even as two Type I crews successfully worked to contain the fires in Frijoles Canyon, a spot fire to the east of the prescribed burn area had grown to 100 acres, and it had spotted an additional quarter mile up the road. In response, fire managers decided to burn sections between State Route 501 and the Camp May Road, Forest Road 1, in an effort to protect the town and the Los Alamos laboratory. Fire conditions were changing and so was the need for response.[20]

That evening, the fire rapidly spiraled out of control. Even though crews contained the Frijoles Canyon spot fire, conditions rapidly worsened. The Type I team took charge of the fire at 6:00 a.m. on May 8, as the fire ran across the east face of the mountain with flames of 100–150 feet in height. On May 9, the *Los Alamos Monitor*, the local newspaper, trumpeted a headline that read: "Wildfire! Worse Fears Become Reality for Los Alamos." As the fire gained momentum, New Mexico governor Gary Johnson ordered the evacuation of Los Alamos, almost 12,000 people. Between 5:00 p.m. on May 10 and early Thursday morning, fires burned across more than 20,000 acres. A total of 239 homes were destroyed in Los Alamos. The fire then moved north in the direction of the San Ildefonso and Santa Clara Pueblo lands adjacent to the Los Alamos installation. Sixty-mile-per-hour winds made the fire devastating and dangerous. Before it was over, it destroyed thirty-nine Los Alamos National Laboratory office trailers and sheds. In a stroke of fortune, no radiation escaped nor was any toxic material released. By the time the fire was brought under control in early June, more than 400 families had been displaced, estimated damage costs exceeded $1 billion, and the idea of prescribed burning faced another enormous challenge.[21]

The Cerro Grande fire was hardly unique in the annals of NPS fire management, but in terms of impact, it was the worst prescribed burn to go awry. While earlier prescribed burns had escaped or caused damage, the scope and scale of Cerro Grande's damage far exceeded any prior escape. Worse, to the public and the press, Cerro Grande looked like a mistake in judgment. Sentiments in the nearby communities became intensely negative. Many openly excoriated the NPS. A few chose not to place blame. Among those who had seen their homes burn either in person or on television, a number remained sanguine about the result. Louis Jalbert, a waste handler at the Los Alamos National Laboratory whose own home and his in-laws' house next door were both destroyed in the fire, felt that the fire was an "act of nature in a tinderbox." Expressing no bitterness as he talked to the *Los Alamos Monitor*, Jalbert still believed that prescribed burns were good policy.[22]

Jalbert held a minority view. Most of the people affected by the fire felt that their trust had been abused, their safety compromised, and their lives put in danger to serve what they regarded as obscure purposes. Their view of the NPS and its fire program was harsh, and even the attempts at apology from the NPS were rebuked. "Based on what we knew at the time and what we believed needed to be done," Superintendent Roy Weaver of Bandelier National Monument told the *Monitor*, he made the decision to start the fire. The results had been devastating, and people "had a right to be frustrated and angry."[23]

The official response came quickly. On May 11, Secretary of the Interior Bruce Babbitt and Secretary of Agriculture Dan Glickman suspended all federal prescribed burning west of the 100th meridian, the line running from North Dakota through Texas, for thirty days. Babbitt formed an interagency fire team to examine the circumstances of the burn. One week later, on May 18, 2000, the investigation report was complete. It was devastating to the NPS fire program. Investigators determined that the prescribed fire plan was not adequate for the circumstances. The complexity rating process for the Cerro Grande area had been flawed. It did not follow the NPS rating system, nor had it been accurately rated. Later reports determined that the fire management rating posted on the Internet when the burn was planned had been incorrect. The danger presented by the conditions at the time of the fire was not adequately understood, in the estimation of the investigation team, nor was interagency cooperation sufficient to assure a useful fire rating system. The investigators recommended that federal agencies should jointly develop interagency complexity rating standards by geographic region rather than try to implement agency-wide standards. The review also determined that the prescribed fire plan had not received a thorough review prior to its approval by Superintendent Roy Weaver, and the prescribed fire planner did not receive sufficient support nor oversight for the task of developing a plan for the prescribed fire.[24]

The press response further deepened the problems for the National Park Service. The juxtaposition of plutonium and other radioactive materials with an intentionally ignited prescribed burn spurred many to the limits of journalistic license. "An Out-of-Control Wildfire: A Nuke Factory with Enough Plutonium to Wipe Out the Entire Southwest: A Handful of Exhausted Firefighters," the headlines in *Maxim* magazine read in an overstated version of a widely asked question. "Just how close did we come to annihilation?"[25]

The Cerro Grande fire represented the first time that critics could point to clear NPS mistakes as the cause of a major fire. Unlike Yellowstone in 1988—when lightning was the genesis of the fires—at Cerro Grande, the NPS simply erred. The prescribed fire had been set in less than optimal conditions; the service lacked both adequate procedures and protocols for fire management; and the timing of the prescribed burn turned out to be poorly chosen. The attendant destruction of homes and Los Alamos National Laboratory structures compounded the problem. Not only did the initial decision reflect poorly on NPS judgment, but the circumstances in which it occurred—near a town of more than 10,000 people and adjacent to the remarkable and potentially devastating research facilities in Los Alamos—made what might only have been an untimely decision look unwise if not downright irresponsible. The fire represented another watershed, one more way that politics and public relations grappled with science as the dominant mode of preparation for and reaction to fire.

Cerro Grande set off more than immediate recriminations. The instantaneous reintroduction of suppression was the beginning of another rethinking of NPS's fire policy. The board of inquiry determined that of the four people with primary responsibility who remained in the National Park Service after the fire, three required more training. Whether intended as an indictment or not, the judgment had the effect of further damning NPS professionalism.[26]

The blame mounted until Bandelier superintendent Roy Weaver spoke out nearly one year after the blaze. Although Weaver had been blamed for the fire—vilified, castigated, and threatened with the loss of his pension in its aftermath—he was quiet until April 2001, when he publicly spoke out in defense of the staff of Bandelier National Monument. The board of inquiry's final report exonerated Weaver, but he believed that the report did not go far enough. In the former superintendent's view, Bandelier had been "unfairly scapegoated" for the fire, he told reporter Keith Easthouse. Not only had the park not been warned of impending high winds on May 7, as had been reported, but the federal report on the fire was so hastily completed that it did not give a fair accounting of the incident or its suppression. "I don't want to deny our responsibility for igniting the prescribed fire," Weaver avowed. "But we did it with a plan that seemed valid and workable. Things happened that we couldn't or didn't anticipate. And that we couldn't control."[27]

This admission was as candid as it was clear. Simply put, fire cannot easily be made to conform to bureaucratic measurements. It is always a risk, always a danger, whether it burns or it is suppressed. All the planning in the world cannot obviate a disastrous change in weather or geographic conditions. Even the board of inquiry, critical in its stance toward the fire response team, recognized the limits of human response. "While the Board did find errors in judgment," the report read, "it also finds that the planning and implementation actions of the principals were not arbitrary, capricious, or unreasonable in light of the information they had prior to the burn and were in compliance with DO-18, RM-18, and other applicable sections of the National Wildland fire policy."[28]

By the middle of the first decade of the twenty-first century, little had been settled about the direction of fire policy in the United States. After Cerro Grande, a series of fires on federal lands further illustrated the problems of the existing regime. During 2002, two "monster fires," in Pyne's words, Hayman and Rodeo-Chediski, were the worst on record in the nearly 100-year history of recordkeeping in Colorado and Arizona. The Biscuit fire in Oregon that same year was easily the state's worst since the nineteenth century. In 2003, southern California's mountains went up in flames. Fires burned across more than 740,000 acres, with twenty-two fatalities and more than 3,000 structures consumed. A new era seemed to dawn, what renowned fire scholar Stephen J. Pyne called "a crash in nature's economy as profound as in the stock market." The terms for fire had to change throughout the federal system, but deciding what would replace the existing structure remained a complex process.[29]

By the twenty-first century, the National Park Service had reached a commanding position as a fire management organization. A century earlier, fire management at places such as Yellowstone National Park had been what the U.S. Cavalry determined it to be—often vain efforts at suppression accomplished with whatever resources were at hand. By 2000, a multifaceted bureaucracy oversaw fire management throughout the scattered dominion of the National Park Service. The NPS participated in interagency fire efforts, keeping a staff of 41 at the National Interagency Fire Center in Boise, Idaho. In 2003, the NPS invested almost $124 million in fire management and another $1.5 billion in Operation of National Park System funding. It employed 434 permanent firefighters, 7 regional fire management officers, 2 Type I hot shot crews, and 1 smoke-jumping crew, as well as 9 fire use modules. The NPS owned 155 fire engines, 14 water/foam tenders, and 9 fire helicopters. This remarkable investment of funds and personnel created a comprehensive fire management program unequaled in the history of the National Park Service.[30]

The change in fire management reflected not only changes in the National Park Service, but those of the twentieth century as well. The goals and standards of early fire suppression had evolved into a management process measured by the highest scientific standards and aimed at achieving goals that were inconceivable at the 1916 inception of the NPS. The army's initial emphasis on suppression in Yellowstone had been replaced by a systematic management structure that reflected improved technologies, better communications, and the cutting edge of ecological science combined with NPS values.

The greatest issues still arose at the intersection of politics and scientific management, when the ideals of policy were not applied with the clarity with which they were conceived or when even the best policy failed to defeat wind-blown fire. The long-standing policy of suppression had made huge tracts of land particularly vulnerable to fires driven by high fuel loads. Such situations became more common as people encroached on land with a propensity to burn even as the NPS and countless other federal agencies scrambled to treat the effects of nearly a century of fire suppression.

Twice, national parks have led a national move to manage fires. In the first instance, when the U.S. Cavalry arrived at Yellowstone, the national parks became the incubator of the idea of national fire management, the place where the experiment to attempt to suppress fire in a systematic way took place. In the second instance, in the late 1960s, the NPS introduced the idea of using fire as a tool, an idea that the Forest Service had buried in its enthusiasm for putting fires out early in the century. Despite the difference in the NPS's mission, its values had spread to its peer agencies and had rewritten the rules of fire management.

By the early twenty-first century, the second heroic age of fire management was passing. The leaders who devised and instituted policies to use fire and then grappled with its consequences began to retire, supplanted by a generation that had never known a complete suppression regime. As the people who had introduced fire to the national parks left the scene, they ceded the ground to this new cadre, who took the prerogative of using fire for granted. This simple change was a manifestation of the triumph of the fire management regime, testimony to its ability to overwhelm the model of suppression that preceded it.

Yet the National Park Service's fire issues remained apart from those of other federal agencies at a time when interagency cooperation was not just desirable but an essential condition for an adequate response to fire. The unique mission of the NPS among federal agencies, its mandate to preserve as well as use the national parks, continued to make the particulars of its fire management more difficult. The service contributed to interagency efforts in the same proportions as did other agencies, but used those resources in different and sometimes more complex ways. Its ability to initiate fires to attempt to recreate images of historical landscapes under the aegis of its resource management program allowed the NPS a measure of flexibility that advocates of the use of fire in other agencies envied.

In a larger setting, this advantage was negated. The western fire scene was "the sum of all we have done and not done over the past century; not only the logging, the grazing, and the road building, but the biosphere reserves, the wilderness areas, the recreational sites; the loss of old species, the invasion of new," Stephen J. Pyne wrote in his 2004 opus, *Tending Fire*. "The fires suppressed, the fires no longer set; the whole rearranged biota of the public domain," he continued. "There is a good case to be made that policy of any sort can not stand under that legacy." Under such circumstances, the success of any fire policy might demand a faith that it does not merit. "Fire's story is not wholly ours to narrate," Pyne reminded his readers, and federal fire managers faced that fact in the early years of the twenty-first century.[31]

For the National Park Service, the dilemma remained: how to get the right fires in the right places and keep the wrong fires out of the wrong areas. More complicated than either all-out suppression or prescribed fire in all its forms as implemented before 2000, this new calculus required even more of the National Park Service than any preceding approach. The service's mission simultaneously complicated its response to fire and shielded it from the sometimes narrow constraints under which other federal agencies functioned. Yet, after Outlet and Cerro Grande, the world was different. After more than a century of dealing with fire in national parks, another new era had begun. In the rest of the twenty-first century, the National Park Service will again have to redefine the boundaries of its fire management strategy.

Notes

INTRODUCTION

1. Horace M. Albright to the Director, August 4, 1926; Horace M. Albright to the Director, August 14, 1926, National Archives and Records Administration (hereafter NARA), Record Group (hereafter RG) 79.7, Glacier National Park, Box 23.

2. Horace M. Albright to the Director, August 14, 1926.

3. Ibid.

4. Park Management Biologist to Chief Park Ranger, Memorandum: Prescribed Burn Experiment, November 18, 1965, Yellowstone National Park, N16.

5. Ibid., 2.

ONE

1. *Report of the Superintendent to the Yellowstone National Park, 1886* (Washington, D.C.: Government Printing Office, 1917), 6–7.

2. H. Duane Hampton, *How the U.S. Cavalry Saved Our National Parks* (Bloomington: Indiana University Press, 1971), 32–33; Alfred Runte, *National Parks: The American Experience*, 2d ed. (Lincoln: University of Nebraska Press, 1987), 35–54; Paul Schullery, *Searching for Yellowstone: Ecology and Wonder in the Last Wilderness* (New York: Houghton Mifflin, 1997); Paul Schullery and Lee Whittlesey, "Yellowstone's Creation Myth: Can We Live with Our Own Legends?" *Montana: The Magazine of Western History* 53, no. 1 (Spring 2003): 2–13.

3. Aubrey L. Haines, *The Yellowstone Story*, vol. 2 (Yellowstone, Wyo.: Yellowstone Library and Museum Association, 1977), 31, 448–449; Hampton, *How the U.S. Cavalry Saved Our National Parks*, 33–35; Runte, *National Parks*, 41–46.

4. *Report of the Superintendent to the Yellowstone National Park, 1879* (Washington, D.C.: Government Printing Office, 1917), 22.

5. *Report of the Superintendent to the Yellowstone National Park, 1882* (Washington, D.C.: Government Printing Office, 1917), 9; Philetus W. Norris to Secretary of the Inte-

rior, June 18, 1878, RG 79, series 3, Correspondence from Yellowstone, 1877– (microfilm), National Archives, College Park, Maryland; Hampton, *How the U.S. Cavalry Saved Our National Parks,* 45–49; Hiram Chittenden, *The Yellowstone National Park* (Cincinnati, Ohio: Clark, 1905), 123–125; Richard A. Bartlett, *Yellowstone: A Wilderness Besieged* (Tucson: University of Arizona Press, 1985), 13–21.

6. Hampton, *How the U.S. Cavalry Saved Our National Parks,* 79–80; Harvey Meyerson, *Nature's Army: When Soldiers Fought for Yosemite* (Lawrence: University Press of Kansas, 2000), 80–81.

7. *Report of the Superintendent to the Yellowstone National Park, 1886,* 6–7.

8. Stephen J. Pyne, *Fire in America: A Cultural History of Wildland and Rural Fire* (Princeton, N.J.: Princeton University Press, 1982), 100–110; Stephen J. Pyne, *Year of the Fires: The Story of the Great Fires of 1910* (New York: Viking, 2001), 112–113.

9. *Report of the Superintendent to the Yellowstone National Park, 1886,* 7; Pyne, *Fire in America,* 118.

10. *Report of the Superintendent to the Yellowstone National Park, 1886,* 8–9; "No Railroad in Yellowstone Park," *Forest and Stream,* February 18, 1886; "Fires in the National Parks," *Forest and Stream,* October 7, 1886.

11. "Putting Out the Fires," *Forest and Stream* 33, no. 1 (July 25, 1889), 1; Robert Utley, *Frontier Regulars: The United States Army and the Indian, 1866–1891* (New York: Macmillan, 1973), 205–213.

12. *Forest and Stream,* February 16, 1886, 62; *Forest and Stream,* October 7, 1886, 1; *Forest and Stream,* October 14, 1886, 226; *Forest and Stream,* July 25, 1889, 1; John F. Reiger, *American Sportsmen and the Origins of Conservation,* rev. ed. (Norman: University of Oklahoma Press, 1986), 32–34, 60–62, 93–142; "Putting Out the Fires," 1.

13. Stephen J. Pyne, *Vestal Fire: An Environmental History, Told through Fire, of Europe and Europe's Encounter with the World* (Seattle: University of Washington Press, 1997), 442–446; Charles S. Sargent, *Report on the Forest of North America (Exclusive of Mexico)* (Washington, D.C.: U.S. Government Printing Office, 1884).

14. *Report of the Superintendent of the Yellowstone National Park, 1888* (Washington, D.C.: Government Printing Office, 1917), 4; *Report of the Superintendent of the Yellowstone National Park, 1889* (Washington, D.C.: Government Printing Office, 1917), 5; *Report of the Superintendent of the Yellowstone National Park, 1890* (Washington, D.C.: Government Printing Office, 1917), 5; Hampton, *How the U.S. Cavalry Saved Our National Parks,* 97–99; Doug Weber, "Fighting Fire with Firepower: Firefighting in Yellowstone National Park, 1872–1918," *Yellowstone Science* 8, no. 3 (Summer 2000): 2–5.

15. *Report of the Superintendent of the Yellowstone National Park, 1892* (Washington, D.C.: Government Printing Office, 1917), 4–5; "Sheepherders and the National Parks," *Forest and Stream,* August 4, 1894.

16. Alfred Runte, Jr., *Yosemite: The Embattled Wilderness* (Lincoln: University of Nebraska Press, 1990), 1–15, makes the cases for Yosemite as the nation's first national park.

17. Steen, *The U.S. Forest Service: A History,* 22–47; G. Michael McCarthy, *Hour of Trial: The Conservation Conflict in Colorado and the West, 1891–1907* (Norman: University

of Oklahoma Press, 1977), 11–17; David A. Clary, *Timber and the Forest Service* (Lawrence: University Press of Kansas, 1986), 3–6. Created under the auspices of Amendment 24 to the General Appropriations Act of 1891, the forest reserves were the predecessors of the national forests.

18. Gifford Pinchot, *Breaking New Ground* (New York: Harcourt Brace, 1947), 44; Pyne, *Fire in America*, 302.

19. T. J. Jackson Lears, *No Place of Grace: Anti-Modernism and the Transformation of American Culture, 1880–1920* (New York: Athenaeum, 1980), 3–7; Stephen J. Pyne, *How the Canyon Became Grand: A Short History* (New York: Viking, 2000), xi–xv, 12–22; Michael P. Cohen, *The Pathless Way: John Muir and American Wilderness* (Madison: University of Wisconsin Press, 1984), 271–272, 288–290.

20. Runte, *Yosemite*, 15–37; Marguerite S. Shaffer, *See America First: Tourism and National Identity, 1880–1940* (Washington, D.C.: Smithsonian Institution), 261–310.

21. C. Kristina Roper Wickstrom, "Issues concerning Native American Use of Fire: A Literature Review," *Yosemite Research Center, Publications in Anthropology*, No. 6 (1987); Mark David Spence, *Dispossessing the Wilderness: Indian Removal and the Making of the National Parks* (New York: Oxford University Press, 1999), 101–108; *Runte, Yosemite*, 10–12, 38.

22. George Gruell, *Fire in Sierra Nevada Forests: A Photographic Interpretation of Ecological Change since 1849* (Missoula, Mont.: Mountain, 2001); Runte, *Yosemite*, 39.

23. Runte, *Yosemite*, 22–24, 49–53; Meyerson, *Nature's Army*, 265.

24. *Report of the Commissioners to Manage the Yosemite Valley and the Mariposa Big Trees, 1889–1890* (Sacramento, Calif.: Superintendent of State Documents, 1890), 6; Spence, *Dispossessing the Wilderness*, 102.

25. *Report of the Commissioners to Manage the Yosemite Valley and the Mariposa Big Trees, 1889–1890*, 7–10. There is some debate about the frequency of fire in the Mariposa Grove. In his November 8, 1890, report to the secretary of the interior, Lieutenant George Davidson notes that the "effects of the fire that swept through the grove in fall of 1888 are painfully apparent." T. W. Swetnam, C. H. Baisan, A. C. Caprio, R. Touchan, and P. M. Brown, *Tree-Ring Reconstruction of Giant Sequoia Fire Regimes* (Final Report on Cooperative Agreement No. DOI 80181-0002, Sequoia and Kings Canyon National Parks, California, 1992), indicates no tree-ring evidence of fire in the grove in 1888. It is possible that Davidson was mistaken in the date for the fire. There are accounts of fire in July 1889, and it may be that Davidson accepted an inaccurate report of when the fires occurred.

26. Michael P. Cohen, *The History of the Sierra Club, 1872–1970* (San Francisco: Sierra Club Books, 1988), 12–14.

27. Runte, *Yosemite*, 52–55; Frank Norris, *The Octopus: A Story of California* (New York: Doubleday, 1901). Members of Muir's group soon became the founding members of the Sierra Club; see Cohen, *History of the Sierra Club*.

28. Cohen, *The Pathless Way*, 288–290.

29. Secretary of the Interior, "Rules for National Parks," RG 79, series 3, Correspondence from Yellowstone, 1877– (microfilm), National Archives (hereafter NA), College Park, Maryland.

30. Thomas Newsham to Secretary of the Interior, November 24, 1890, NARA, RG 79, Records of the Office of the Secretary of the Interior Relating to National Parks, Box 89.

31. Lt. George Davidson, "Report on the Alleged Spoilations [*sic*] in the Yosemite Valley," November 8, 1890, RG 79, Records of the Office of the Secretary of the Interior Relating to National Parks, Box 89; *Reports of the Secretary of the Interior Relative to Yosemite Park, 1892* (Washington, D.C.: Government Printing Office, 1893), 4; *Report of the Acting Superintendent of the Yosemite National Park to the Secretary of the Interior, 1891* (Washington, D.C.: Government Printing Office, 1891), 8.

32. Meyerson, *Nature's Army*, 93–97.

33. *Report of the Commissioners to Manage the Yosemite Valley and the Mariposa Big Trees, 1889–1890*, 6–7; Meyerson, *Nature's Army*, 97–98.

34. *Reports of the Secretary of the Interior Relative to Yosemite Park, 1892*, 7–10.

35. Pyne, *Fire in America*, 102.

36. U.S. Attorney General Richard Olney to Secretary of the Interior Hoke Smith, September 28, 1894, RG 79, Records of the Office of the Secretary of the Interior Relating to National Parks, Box 89.

37. "Brief Report of Captain J. H. Dorst, Acting Superintendent of Sequoia National Park," ca. 1891, RG 79, Letters Received by Office of the Secretary of the Interior Relating to National Parks, Sequoia and General Grant 1890–1907, Box 47; Lary M. Dilsaver and William C. Tweed, *The Challenge of the Big Trees: A Resource History of Sequoia and Kings Canyon National Parks* (Three Rivers, Calif.: Sequoia Natural History Association, 1990), 66–72.

38. J. H. Dorst to Secretary of the Interior, August 4, 1892, RG 79, Letters Received by Office of the Secretary of the Interior Relating to National Parks, Sequoia and General Grant 1890–1907, Box 48.

39. E. Louise Peffer, *The Closing of the Public Domain: Disposal and Reservation Policies 1900–1950* (Palo Alto, Calif.: Stanford University Press, 1950), 8–31, 43–44.

40. W. F. Landers, "Report on the Investigation of Causes and Effects of Forest Fires in California," September 1894, RG 79, Letters Received by Office of the Secretary of the Interior Relating to National Parks, Sequoia and General Grant 1890–1907, Box 49.

41. J. W. Zaveley to Secretary of the Interior, August 4, 1898; Henry B. Clark to Secretary of the Interior, September 30, 1899, RG 79, Letters Received by Office of the Secretary of the Interior Relating to National Parks, Sequoia and General Grant 1890–1907, Box 49.

42. J. W. Zaveley to Secretary of the Interior, August 4, 1898; Henry B. Clark to Secretary of the Interior, September 30, 1899; Henry B. Clark to Secretary of the Interior, October 31, 1899, RG 79, Letters Received by Office of the Secretary of the Interior Relating to National Parks, Sequoia and General Grant 1890–1907, Box 49.

43. Pyne, *Year of the Fires*, 2–3.

44. Superintendent Capt. George S. Anderson, "Work of the Cavalry in Protecting the Yellowstone National Park," *Journal of the United States Cavalry Association* 10, no. 36 (March 1897): 6–7; *Report of the Acting Superintendent of Yellowstone Park, 1901* (Washington, D.C.: Government Printing Office, 1902), 4; "Fires in Yellowstone," *Forest and Stream*, August 10, 1901, 102.

45. Chittenden, *Yellowstone National Park,* 242–244.

46. Pyne, *Year of the Fires,* 116; "Fires in Yellowstone Park," *Forest and Stream,* September 24, 1910, 494.

47. Telegram, Clements Ucker to Franklin Pierce, August 10, 1910; Telegram, Clements Ucker to Franklin Pierce, August 12, 1910, RG 79, series 7, Glacier National Park, General Records, Expenditures/Supplies/Materials/Fires, Box 22; Pyne, *Year of the Fires,* 109–111; Spence, *Dispossessing the Wilderness,* 77–90.

48. Hal K. Rothman, ed., *"I'll Never Fight Fire with My Bare Hands Again": Recollections of the First Forest Rangers of the Inland Northwest* (Lawrence: University Press of Kansas, 1994), 66–89; Pyne, *Year of the Fires,* 201, 233.

49. Pyne, *Fire in America,* 243–245.

50. Telegram, Franklin Pierce to Clements Ucker, August 15, 1910, RG 79, series 7, Glacier National Park, Expenditures/Supplies/Materials/Fires, Box 22; Pyne, *Year of the Fires,* 111.

51. Superintendent, Glacier National Park to Secretary of the Interior, October 9, 1910, RG 79, series 7, Glacier National Park, Expenditures/Supplies/Materials/Fires, Box 22; Pyne, *Year of the Fires,* 117–122; Pyne, *Fire in America,* 244.

52. Benson to Secretary of the Interior, October 27, 1910, Yellowstone Box Y-9, Letter Box 65, Yellowstone National Park Library; "Fire in Yellowstone," *Forest and Stream,* September 24, 1910, 494.

53. Franklin Pierce to Major W. R. Logan, September 7, 1910; W. R. Logan to Secretary of the Interior, September 14, 1910, RG 79, series 7, Glacier National Park, Expenditures/Supplies/Materials/Fires, Box 22; Pyne, *Fire in America,* 244.

54. "Forest Fires of the Season of 1910 in National Parks," RG 79, series 7, Glacier National Park, Expenditures/Supplies/Materials/Fires, Box 22.

55. Pyne, *Fire in America,* 100–103.

56. Pyne, *Fire in America,* 104; G. L. Hoxie, "How Fire Helps Forestry," *Sunset* 25 (August 1910): 145–151.

57. Pyne, *Fire in America,* 104–105.

58. Secretary of War to Secretary of the Interior, May 1, 1914, RG 79, Records of the Office of the Secretary of the Interior Relating to National Parks, Box 89; Robert Shankland, *Steve Mather of the National Parks* (New York: Knopf, 1951), 104–113.

TWO

1. Robert Shankland, *Steve Mather of the National Parks,* 2nd ed. (New York: Knopf, 1970), 1–41.

2. Ibid., 7, 53–54; Alfred Runte, *National Parks: The American Experience,* 2nd ed. (Lincoln: University of Nebraska Press, 1987), 109–110.

3. Hal K. Rothman, *Preserving Different Pasts: The American National Monuments* (Urbana: University of Illinois Press, 1989); Runte, *National Parks;* An Act for the Preservation of American Antiquities, 16 USC 431–433, June 8, 1906.

4. Samuel P. Hays, *Conservation and the Gospel of Efficiency: The Progressive Conservation Movement, 1880–1920* (Cambridge, Mass.: Harvard University Press, 1959), 1–5; Ste-

phen J. Pyne, *Fire in America: A Cultural History of Wildland and Rural Fire in the United States* (Princeton, N.J.: Princeton University Press, 1982), 295–296; Mather's official files in the National Archives, RG 79, for this period show no mention of fire.

5. Pyne, *Fire in America*, 295–297; Hal K. Rothman, ed., *"I'll Never Fight Fire with My Bare Hands Again": Recollections of the First Forest Rangers of the Inland Northwest* (Lawrence: University Press of Kansas, 1994), 1–16, 110–136; Harold K. Steen, *The U.S. Forest Service: A History* (Seattle: University of Washington Press, 1976), 37–59.

6. Pyne, *Fire in America*, 296–297; David Carle, *Burning Questions: America's Fight with Nature's Fire* (Westport, Conn.: Praeger, 2002), 11–26; Runte, *National Parks*, 102–104; Shankland, *Steve Mather of the National Parks*, 92–99, 133–142.

7. Runte, *National Parks.*

8. Carle, *Burning Questions*, 236–238; Lary M. Dilsaver and William C. Tweed, *Challenge of the Big Trees: A Resource History of Sequoia and Kings Canyon National Parks* (Three Rivers, Calif.: Sequoia Natural History Association, 1990), 151–169.

9. Rothman, *"I'll Never Fight Fire with My Bare Hands Again,"* 103–136.

10. Pyne, *Fire in America*, 353–356; Carle, *Burning Questions*, 238; Steen, *The U.S. Forest Service: A History*, 129–131, 176–179.

11. A thorough review of agency files in NA, RG 79, series 6, Central Files, and series 7, Central Classified Files, reveals no examples of proactive fire planning. Albright's two memoirs, Albright and Robert Cahn, *The Birth of the National Park Service: The Founding Years, 1913–1933* (Salt Lake City, Utah: Howe, 1985), and Albright and Marian Schenck, *Creating the National Park Service: The Missing Years* (Norman: University of Oklahoma Press, 1999), contain a total of five references to fire, all but one of which are comments on the impact or aftermath of specific fires. No correspondence of Mather in the National Archives, RG 79, suggests any proactive approach to fire. The small collection of his papers at the Bancroft Library at the University of California, Berkeley, only peripherally addresses the early years of the NPS. Although it is difficult to prove a negative, the author is quite convinced that the National Park Service had no comprehensive fire strategy in its early years.

12. Department of the Interior, *Report of the Director of the National Park Service to the Secretary of the Interior, for the Fiscal Year Ended June 30, 1917* (Washington, D.C.: Government Printing Office, 1917), 34.

13. *Superintendent's Annual Report, Glacier National Park, 1911*, 7; *Superintendent's Annual Report, 1916*, 8, National Archives, RG 79, Glacier National Park, series 6, Superintendents' Annual Reports, 1910–1983.

14. *Superintendent's Annual Report, Glacier National Park, 1919*, 22; W. W. Payne to the Director, December 30, 1919; George Goodwin to Director, September 9, 1920, National Archives, RG 79, Glacier National Park, series 6, Central Files, 1907–1939, Box 22; Mark D. Spence, *Dispossessing the Wilderness: Indian Removal and the Making of the National Parks* (New York: Oxford University Press, 1999), 88–95.

15. W. W. Payne to Director, May 17, 1920, Glacier National Park, series 6, Central Files, 1907–1939, Box 22.

16. Ibid.

17. *Superintendent's Annual Report, Glacier National Park, 1918*, 35; *Superintendent's Annual Report, Glacier National Park, 1919*, 22; *Superintendent's Annual Report, Glacier National Park, 1923*, 8, all in Glacier National Park Archives; Hal K. Rothman, "'A Regular Ding-Dong Fight': Agency Culture and Evolution in the Park Service–Forest Service Dispute, 1916–1937," *Western Historical Quarterly* 20, no. 2 (May 1989): 141–161.

18. *Superintendent's Annual Report, Glacier National Park, 1921*, 21; *Superintendent's Annual Report, Glacier National Park, 1923*, 15; *Superintendent's Annual Report, Glacier National Park, 1925*, 11, all in Glacier National Park Archives.

19. Pyne, *Fire in America*, 266–272.

20. *Fourth Annual Report of the National Park Service, 1920: S. T. Mather, Director* (Washington, D.C.: Government Printing Office, 1920), 278–301; *Report of the Director of the National Park Service for the Fiscal Year Ended June 30, 1927* (Washington, D.C.: Government Printing Office, 1928), 104–117; Barry Mackintosh, *Visitor Fees in the National Park System: A Legislative and Administrative History* (Washington, D.C.: National Park Service, 1983), 2–10; John Ise, *Our National Park Policy: A Critical History* (Baltimore, Md.: Johns Hopkins University Press, 1961), 619–624.

21. Arno B. Cammerer to Secretary of the Interior, September 15, 1922; Arno B. Cammerer to Secretary of the Interior, June 11, 1924, Glacier National Park, series 6, Central Files, 1907–1939, Box 23.

22. *Superintendent's Annual Report, Glacier National Park, 1924*, 8.

23. Arno B. Cammerer to Colonel White, September 8, 1924; Horace M. Albright to Sir, September 15, 1924; John R. White to Arno B. Cammerer, September 20, 1924, all in Fire Records, 1904–1930 Man-U Box 275, Folder 2, Sequoia National Park.

24. *Superintendent's Annual Report, Sequoia National Park, 1926*, 10.

25. Act of June 7, 1924, PL 68-270, Ch. 348, 43 Stat. 653; Laws and Legislation, 1924: Clarke-McNary Act, U.S. Forest Service Headquarters Collection, Forest History Society, Durham, N.C.; William G. Robbins, *A History of National, State, and Private Cooperation* (Lincoln: University of Nebraska Press, 1985), 85–104; Pyne, *Fire in America*, 353–357.

26. Shankland, *Steve Mather of the National Parks*, 262; Donald W. Swain, *Wilderness Defender: Horace M. Albright and American Conservation* (Chicago: University of Chicago Press, 1970); William C. Winkler and Merrie H. Winkler, "Ansel F. Hall, 1894–1962," in William Sontag, ed., *National Park Service: The First 75 Years* (New York: Eastern National Park and Monument Association, 1991), 21–22.

27. John R. White to Arno B. Cammerer, September 30, 1924, Fire Records, 1904–1930, Man-U Box 275, Folder 2, Sequoia National Park.; *Superintendent's Annual Report, Sequoia National Park, 1925*, 4; Charles Kraebel to Arno B. Cammerer, September 3, 1924, NA, RG 79, Central Files, 1907–1939, Glacier National Park, Box 23.

28. Charles Kraebel to the Director, telegram, August 15, 1925, NA, RG 79, Central Files, 1907–1939, Glacier National Park, Box 23.

29. Charles Kraebel to George Slack, June 6, 1926, NA, RG 79, Central Files, 1907–1939, Glacier National Park, Box 23; Arno B. Cammerer to Stephen T. Mather, July 31,

1926; Department of the Interior, Memorandum for the Files: For Immediate Release, August 3, 1926; Arno B. Cammerer to H. A. Noble, August 3, 1926; Horace M. Albright to the Director, August 4, 1926, NA, RG 79, Central Files, 1907–1939, Glacier National Park, Box 23.

30. Horace M. Albright to the Director, August 4, 1926; Horace M. Albright to the Director, August 14, 1926, NA, RG 79, Central Files, 1907–1939, Glacier National Park, Box 23.

31. Horace M. Albright to the Director, August 14, 1926.

32. Ibid.

33. "Fire History of Glacier National Park (Based on Extracts from Glacier National Park Master Plan and Other Records)," NA, RG 79, Central Files, 1907–1939, Glacier National Park, Box 23.

34. Winkler and Winkler, "Ansel F. Hall," 21.

35. Shankland, *Steve Mather of the National Parks*, 259–262; Swain, *Wilderness Defender*; Winkler and Winkler, "Ansel F. Hall," 21; Pyne, *Fire in America*, 298.

36. Horace M. Albright to M. B. Pratt, November 1, 1928; Superintendent, Sequoia National Park to H. M. Albright, November 8, 1928; J. D. Coffman, Memorandum Re: Colonel White's Letter of November 8, November 12, 1928, Fire Records, 1904–1930, Box F2, Sequoia National Park.

37. Shankland, *Steve Mather of the National Parks*, 262; Pyne, *Fire in America*, 112; Carle, *Burning Questions*, 135–137.

38. Yearly Report of Superintendent (Col. John R. White), 1926, Man-L, Box 3, Sequoia National Park Archives; Pyne, *Fire in America*, 270, 298.

39. "Fire History of Glacier National Park (Based on Extracts from Glacier National Park Master Plan and Other Records)," NA, RG 79, Central Files, 1907–1939, Glacier National Park, Box 23.

40. Pyne, *Fire in America*; Coert duBois, *Systematic Fire Protection in the California Forests* (San Francisco, Calif.: USDA Forest Service, Pacific Southwest Region, 1914).

41. John D. Coffman, interview by Herbert Evison, October 29, 1962, Herbert Evison Papers, Western History Collection, Denver Public Library, Denver, Colorado.

42. "Glacier National Park Fire Control Plan: Instructions for the Fire Protection Organization," NA, RG 79, Central Classified Files, 1933–1949, National Parks, Glacier, Box 976.

43. John Coffman interview, October 29, 1962.

44. *Superintendent's Annual Report, 1929*, 1; "Fire History of Glacier National Park: Based on Extracts from Glacier National Park Master Plan and Other Records," 2; Horace M. Albright to Walter Newton, October 25, 1929, NA, RG 79, Central Classified Files, 1907–1949, Box 254.

45. John D. Coffman, "Review of Half Moon Fire—1929," NA, RG 79, Central Classified Files, 1907–1949, Glacier National Park, Box 254, 1–17.

46. Ibid., 18–20; J. Ross Eakin to the Director, October 12, 1929, NA, RG 79, Central Classified Files, 1907–1949, Glacier National Park, Box 254.

47. Swain, *Wilderness Defender*, 206–208.

THREE

1. John D. Coffman, "Report on Fire Protection Requirements of Yellowstone National Park, 1929," Yellowstone Box Y-223, Yellowstone National Park Archives.

2. Roger W. Toll, *Annual Report for Yellowstone National Park 1931*, Yellowstone Box Y-223, Yellowstone National Park Archives, 1.

3. Ibid., 1; John Coffman interview by Herbert Evison, October 28, 1962, Denver Public Library, Western History Collection, NPS Conservation 56, Box 1, Folder C, 6, 13; John D. Coffman, "Report on Fire Protection Requirements of Yellowstone National Park," January 18, 1930, Yellowstone Box Y-223, Yellowstone National Park Archives.

4. Roger W. Toll, *Annual Report for Yellowstone National Park 1931*, 2–3.

5. *Annual Report of the Superintendent of Yellowstone National Park, 1932*, 3–4.

6. Horace M. Albright, *Origins of National Park Service Administration of Historic Sites* (Philadelphia: Eastern National Parks and Monuments Association, 1971), 1–17; Ethan Carr, *Wilderness by Design: Landscape Architecture and the National Park Service* (Lincoln: University of Nebraska Press, 1998), 55–94.

7. Coffman interview, October 29, 1962, 14; Donald Swain, *Wilderness Defender: Horace M. Albright and American Conservation* (Chicago: University of Chicago Press, 1970), 184–188.

8. Richard Polenberg, *Reorganizing Roosevelt's Government: The Controversy over Executive Reorganization, 1936–1939* (Cambridge, Mass.: Harvard University Press, 1966), 1–11; Robert M. Crunden, *A Brief History of American Culture* (New York: Paragon, 1994), 129–159, 185–236; T. J. Jackson Lears, *No Place of Grace: Anti-Modernism and the Transformation of American Culture, 1880–1920*, 2d ed. (Chicago: University of Chicago Press, 1994), 1–58.

9. Hal K. Rothman, *Saving the Planet: The American Response to the Environment in the Twentieth Century* (Chicago: Dee, 2000), 60–84.

10. John C. Paige, *The National Park Service and the New Deal* (Washington, D.C.: National Park Service, 1985), 8–18; Cornelius Maher, *Nature's New Deal* (New York: Oxford University Press, forthcoming).

11. Leuchtenburg, *Franklin D. Roosevelt and the New Deal* (New York: Harper and Row, 1963), 6–9; Rexford Tugwell, *FDR: Architect of an Era* (New York: Macmillan, 1967), i–xiv; Robert A. Caro, *The Years of Lyndon Johnson: The Path to Power* (New York: Knopf, 1982), 241–260; Phoebe Cutler, *The Public Landscape of the New Deal* (New Haven, Conn.: Yale University Press, 1985), 1–8, 90–106.

12. Paige, *The National Park Service and the New Deal*, 21–25.

13. Horace Albright as told to Robert Cahn, *The Birth of the National Park Service: The Founding Years, 1913–1933* (Salt Lake City, Utah: Howe, 1985), 283–288; Tom H. Watkins, *Righteous Pilgrim: The Life and Times of Harold L. Ickes, 1874–1952* (New York: Holt, 1990), 550–555.

14. Albright and Cahn, *Birth of the National Park Service*, 288–290; Swain, *Wilderness Defender*, 219–225.

15. Coffman interview, October 28, 1962; Albright and Cahn, *Birth of the National Park Service*, 292–297; Swain, *Wilderness Defender*, 219–225; Caro, *Years of Lyndon Johnson*, 341–351.

16. Coffman interview, October 28, 1962, 13–14.

17. Ibid., 14; Watkins, *Righteous Pilgrim*, 447–594.

18. Leuchtenburg, *Franklin D. Roosevelt and the New* Deal, 11–15; Paige, *The National Park Service and the New Deal*, 11–19, 213.

19. Paige, *The National Park Service and the New Deal*, 18–19, 213.

20. L. F. Cook, "Forest Fire Protection in the National Park System, 1930–39," Occasional Forestry Note No. 5, March 25, 1940, Sequoia National Park, SNP 42, Box 375, F2, 5–6.

21. Emergency Conservation Work Press Release, October 23, 1933, NARA, RG 79.4.3, Records of Branch of Forestry; Third Report of the Director of Emergency Conservation Work, for the Period April 1934 to September 30, 1934 (with Certain Data from April 5, 1933 through September 30, 1934), 29–46.

22. Robert Shankland, *Steve Mather of the National Parks* (New York: Knopf, 1951), 303.

23. Coffman interview, October 28, 1962, 15–16; National Park Service, "Conference of Superintendents and Field Officers, Washington, D.C., November 19–23, 1934," 145; Harold L. Ickes, *The Secret Diary: The First Thousand Days, 1933–1936* (New York: Simon and Schuster, 1954), 18–19; Shankland, *Steve Mather of the National Parks*, 297–303; Swain, *Wilderness Defender*, 223.

24. Richard West Sellars, *Preserving Nature in the National Parks: A History* (New Haven, Conn.: Yale University Press, 1997), 126–127; Coffman interview, October 29, 1962, 16; J. F. Kieley, *A Brief History of the National Park Service* (Washington, D.C.: Civilian Conservation Corps, Department of the Interior, 1940); Harlan D. Unrau and G. Frank Williss, *Administrative History: Expansion of the National Park Service in the 1930s* (Denver, Colo.: Denver Service Center, 1983), 245.

25. Stephen J. Pyne, *Fire in America: A Cultural History of Wildland and Rural Fire in the United States* (Princeton, N.J.: Princeton University Press, 1982), 270–273, 346–350; David Carle, *Burning Questions: America's Fight with Nature's Fire* (Westport, Conn.: Praeger, 2002), 52.

26. Elers Koch, "The Passing of the Lolo Trail," *Journal of Forestry* 33, no. 2 (1935): 98–104; Elers Koch, *Forty Years a Forester, 1903–1943* (Missoula, Mont.: Mountain, 1998), 91–107; Pyne, *Fire in America*, 278–282; Stephen Pyne, *Year of the Fires: The Story of the Great Fires of 1910* (New York: Viking, 2001), 264–266.

27. Pyne, *Fire in America*, 272–275.

28. *Directory of CCC Camps Supervised by the National Park Service (Updated to December 31, 1941)* (Washington, D.C.: Government Printing Office, 1942), 1, 4, 7.

29. *Annual Report of the Superintendent of Yellowstone National Park, 1933*, 3; *Annual Report of the Superintendent of Yellowstone National Park, 1934*, 4–5; *Annual Report of the Superintendent of Yellowstone National Park, 1935*, 5; Circular No. 25, October 8, 1935; Superintendent's Monthly Report, October 1935, all in Yellowstone National Park Library.

30. *Annual Report of the Superintendent of Yellowstone National Park, 1933*, 3; *Annual Report of the Superintendent of Yellowstone National Park, 1934*, 4–5; *Annual Report of the Superintendent of Yellowstone National Park, 1935*, 5.

31. E. T. Scoyen to the Director, October 11, 1934, NA, RG 79, Glacier National Park, series 7.

32. E. T. Scoyen to Director, September 10, 1936, NA, RG 79, Glacier National Park, series 7.

33. Ibid.; Howard H. Hays to Arno B. Cammerer, September 7, 1936, NA, RG 79, Glacier National Park, series 7; J. D. Coffman, Memorandum for the Superintendent, Sequoia National Park, August 6, 1933, Sequoia National Park, F275, F2.

34. Cook, "Forest Fire Protection in the National Park System, 1930–39," 3; Durwood Dunn, *Cades Cove: The Life and Death of an Appalachian Community, 1818–1937* (Knoxville: University of Tennessee Press, 1989); Warren Hofstra, *The Planting of New Virginia: Settlement and Landscape in the Shenandoah Valley* (Baltimore, Md.: Johns Hopkins University Press, 2004); Margaret L. Brown, *The Wild East: A Biography of the Great Smoky Mountains* (Gainesville: University of Florida Press, 2000).

35. "Report on Mammoth Cave, 1937," NA, RG 79, Branch of Forestry, Forest Fire Reports 1928–1949, Box 2, Entry 33; John Ise, *Our National Park Policy: A Critical History* (Baltimore, Md.: Johns Hopkins University Press, 1961), 254–256.

36. Robert Holland Memorandum, March 26, 1935; John D. Coffman, Memorandum, April 17, 1935, NA, RG 79, Branch of Forestry, Forest Fire Reports 1928–1949, Box 2, Entry 33.

37. Cook, "Forest Fire Protection in the National Park System, 1930–39," 3–4.

38. *Annual Report of the Superintendent of Mesa Verde National Park, 1934*, NA, RG 79, series 7, Mesa Verde, Box 1; Paul Rogers, "Mesa Verde Fire History, March 2002," 5–7, unpublished paper, Mesa Verde National Park Archives, Mesa Verde National Park.

39. Rogers, "Mesa Verde Fire History, March 2002," 7.

40. Adolph Murie to Ben Thompson, July 13, 1935; Adolph Murie, "Memorandum for Ben H. Thompson, August 2, 1935," Entry 34, RG 79, NA, Glacier National Park, Box 973; Paul S. Sutter, *Driven Wild: How the Fight against Automobiles Launched the Modern Wilderness Movement* (Seattle: University of Washington Press, 2002), 23–53; Sellars, *Preserving Nature in the National Parks*, 128–129.

41. Lawrence F. Cook, "Memorandum for the Chief Forester, August 28, 1935," Entry 34, RG 79, NA; Sellars, *Preserving Nature in the National Parks*, 130.

42. Pyne, *Fire in America*, 255–256; Pyne, *Year of the Fires*, 253–267.

43. Sellars, *Preserving Nature in the National Parks*, 126–128.

44. L. F. Cook, "Forest Fire Protection in the National Park System, 1930–1939," Occasional Forestry Note No. 5, March 25, 1940, Department of the Interior, National Park Service, Sequoia National Park Archive, SNP 42, Box 275, F2.

45. Coffman interview, October 28, 1962, 19.

46. Miscellaneous Circular No. 15 for Western Parks and Monuments, December 17, 1934, Sequoia National Park, FR 31–34, Man-U, Box 2, F20, 1–2.

47. Ibid., 4–5.

48. Ibid., 2; *Superintendent's Annual Report, 1940*; John D. Coffman, Memo to Regional Director, Region II, November 8, 1940, Yellowstone National Park, 883-03.3, Y-240.

49. "Grand Canyon National Park, Fire Control Plan, Season of 1939," 1, Grand Canyon National Park," Grand Canyon National Park, Doc. 12; "Fire Control Plan for Grand Teton National Park, 1939," RG 79, Grand Teton National Park, series 7, 100, pt. 1, 2.

50. "Fire Control Plan for Grand Teton National Park, 1939," 4–5; "Grand Canyon National Park, Fire Control Plan, Season of 1939," 2–3.

51. "Fire Cooperation Agreement between Yellowstone National Park and Absaroka National Forest, June 13, 1932," Yellowstone National Park, W-238.

52. R. R. Vincent to Director, August 11, 1936; "Cooperative Agreement," ca. 1936, NA, RG 79, Glacier National Park, series 7, 100, pt. 1; Memorandum to Field Officers A-17-1, Cooperative Fire Fighting Agreements, December 17, 1938, NA, RG 79, Branch of Forestry, Box 281; "Fire Cooperative Agreement: Kaibab National Forest Service–Grand Canyon National Park," Grand Canyon National Park, Doc. 13.

53. Cook, "Forest Fire Protection in the National Park System, 1930–39," 9; Arno B. Cammerer, director, "The National Park Service," in U.S. Department of the Interior, *Annual Report of the Secretary, 1940* (Washington, D.C.: Government Printing Office, 1940), 191–193.

54. Cook, "Forest Fire Protection in the National Park System, 1930–39," 10.

55. Department of the Interior, *Annual Report for the Fiscal Year Ending June 30, 1939* (Washington, D.C.: Government Printing Office, 1939), 273–274, 296; Newton B. Drury, director, "Report of the Director of the National Park Service, 1943," in Department of the Interior, *Annual Report of the Secretary of the Interior, 1944* (Washington, D.C.: Government Printing Office, 1944), 207–208; John C. Paige, *The Civilian Conservation Corps and the National Park Service, 1933–1942* (Washington, D.C.: National Park Service, 1985), 132; Lary M. Dilsaver and William C. Tweed, *Challenge of the Big Trees: A Resource History of Sequoia and Kings Canyon National Parks* (Three Rivers, Calif.: Sequoia Natural History Association, 1990), 159–167.

56. Newton B. Drury, director, "Report of the Director of the National Park Service, 1943," in Department of the Interior, *Annual Report of the Secretary, 1943* (Washington, D.C.: Government Printing Office, 1943), 207–208.

57. Newton B. Drury, "The National Parks in Wartime," *American Forests* (August 1943): 37–42; Drury, "Report of the Director of the National Park Service, 1943," 207.

58. Memorandum for the Director, January 24, 1942, Grand Canyon National Park Archive; Newton B. Drury, director, "National Park Service," in *Report of the Secretary of the Interior, 1944*, 223.

59. "Special Report: Law Enforcement Conference, Forest Service Remount Depot, Huson, Montana, March 3 to March 12, 1942," Yellowstone National Park, W-1; Carl G. Kruger to Superintendent, Yellowstone National Park, April 9, 1943, Yellowstone, Y-232; J. S. Veeder to Colonel R. G. Walters, July 24, 1943; Hugh H. Miller, "Memorandum for the Assistant to the Secretary in Charge of Land Utilization, March 7, 1945," both in Yellowstone, Y-232; Lawrence C. Merriam, "Memorandum for the Superintendent, Yellowstone National Park, April 7, 1943," Yellowstone, Y-230.

60. Memorandum for All Region Four Areas, February 27, 1946, Sequoia National Park Archives, Box 275, F2; "Special Report: Law Enforcement Conference, 1942," 1.

FOUR

1. Board of Review Report, Rancheria Mountain Fire, Yosemite National Park, September 9–21, 1948, 1–3; Yosemite National Park Annual Forestry Report, 1948, 3; Fire Control Plan, January 1949, Yosemite National Park, 1, all in Fire Records by Year, 1931–1974, Yosemite National Park Archives.

2. Board of Review Report, Rancheria Mountain Fire, Yosemite National Park, September 9–21, 1948, 3–4.

3. Ibid.

4. Eric Goldman, *The Crucial Decade—and After: America, 1945–1960* (New York: Random House, 1960), 4–5, 12–15; James T. Patterson, *Grand Expectations: The United States, 1945–1974* (New York: Oxford University Press, 1996), 61–65.

5. Hal K. Rothman, *Devil's Bargains: Tourism in the Twentieth Century American West* (Lawrence: University Press of Kansas, 1998), 202–205.

6. *Superintendent's Annual Report for Sequoia National Park, 1947*, Sequoia National Park, Superintendent's Files, 2; Bernard DeVoto, "Let's Close the National Parks," *Harper's Magazine* 207 (1953): 49–52; David Clary, *Timber and the Forest Service* (Lawrence: University Press of Kansas, 1986); John Jakle, *The Tourist: Travel in Twentieth-Century North America* (Lincoln: University of Nebraska Press, 1985), 185–198; Paul S. Sutter, *Driven Wild: How the Fight against Automobiles Launched the Modern Wilderness Movement* (Seattle: University of Washington Press, 2002).

7. John D. Coffman, "Forest Protection in the National Parks," interview by Amelia R. Fry, 75–77, 1973, Bancroft Library, Berkeley, California.

8. Stephen J. Pyne, *Fire in America: A Cultural History of Wildland and Rural Fire* (Princeton, N.J.: Princeton University Press, 1982); Robert Righter, *Crucible for Conservation: The Struggle for Grand Teton National Park* (Niwot: Colorado Associated University Press, 1982), 103–126; Hal K. Rothman, *Preserving Different Pasts: The American National Monuments* (Urbana: University of Illinois Press, 1989), 225–249.

9. Pyne, *Fire in America*, 371–373.

10. Clarence Strong to Lawrence C. Merriam, March 8, 1950, Yellowstone Archives, Box Y-232; Norman Mclean, *Young Men and Fire* (Chicago: University of Chicago Press, 1992); *Superintendent's Annual Report for Yellowstone, 1950*, Yellowstone National Park.

11. Forester to Regional Forester, February 17, 1950, Yellowstone Archives, Box Y-240.

12. *Superintendent's Annual Report for 1951, Yellowstone National Park*, Yellowstone National Park, Superintendent's Files, 16; Edmund Rogers to Supervisor, Targhee National Forest, July 10, 1950; Edmund Rogers to Julius Schoener, May 20, 1953, Yellowstone N-33; "Smokejumpers," Yellowstone Archives, Box Y-232.

13. Stephen J. Pyne to Hal Rothman, e-mail, January 26, 2004.

14. Fire Control Plan, January 1949, 2–4.

15. National Park Service, Annual Fire Report, January 1–December 31, 1953, 1–6; Acting Regional Director to Region Two Field Areas, February 26, 1954; National Park Service, Annual Fire Report, January 1–December 31, 1954, 1–3; Regional Chief of Operations to Superintendents, Region Two Areas, March 14, 1955, Office of the Superintendent, Grand Teton National Park, Grand Teton National Park Archives.

16. Stephen J. Pyne to Hal Rothman, January 29, 2004, e-mail in possession of the author.

17. National Park Service, Annual Fire Report, January 1–December 31, 1954; Acting Regional Director to Superintendents, Region Two Field Areas, April 10, 1956, Office of the Superintendent, Grand Teton National Park, Grand Teton National Park Archives.

18. Annual Forest Fire Report of the National Park Service, 1956, 1; Regional Director to Superintendents, Region Two Field Areas, March 22, 1957, Office of the Superintendent, Grand Teton National Park, Grand Teton National Park Archives.

19. United States Department of the Interior, National Park Service, *Fire Control Handbook*, ca. 1958, NPS Technical Information Center, Denver, Colorado, microfilm.

20. "Report of the Superintendent of Everglades National Park, 1947," NA, RG 79, Central Files, 1907–1939, Everglades National Park.

21. Everglades National Park Authorization Act, 16 USC, sec. 410 (1948) (enacted 1934). On December 6, 1947, President Harry S Truman dedicated the park with the words:

> Not often in these demanding days are we able to lay aside the problems of the time, and turn to a project whose great value lies in the enrichment of the human spirit. Today we mark the achievement of another great conservation victory. We have permanently safeguarded an irreplaceable primitive area. We have assembled to dedicate to the use of all people for all time, the Everglades National Park.

David McCally, *The Everglades: An Environmental History* (Gainesville: University of Florida, 1999), 1–2.

22. William B. Robertson, *A Survey of the Effects of Fire in Everglades National Park* (Washington, D.C.: National Park Service, 1953), 1.

23. David McCally, *The Everglades: An Environmental History* (Gainesville: University of Florida, 1999), 18–20.

24. Robertson, *A Survey of the Effects of Fire in Everglades National Park*, 3–13; Pyne, *Fire in America*, 302–303; McCally, *The Everglades*, 31–57.

25. Fire Critique, Everglades National Park, May 16, 1950, Records of Key Officials, Box 7, File 19, E-Fees, Everglades National Park Archives.

26. Ibid., 3–4.

27. Robertson, *A Survey of the Effects of Fire in Everglades National Park*, 169; Fire Critique, Everglades National Park, May 16, 1950; Thomas J. Allen to District Engineer, Jacksonville District, U.S. Army Corps of Engineers, August 5, 1949; Devereux Butcher to Newton B. Drury, February 24, 1948; Newton B. Drury to Devereux Butcher, February 27, 1949, Records of Key Officials, Box 7, File 19, E-Fees, Everglades National Park Archives.

28. Daniel Beard, Memorandum to Regional Director, July 11, 1956, Everglades National Park, 883.01; William B. Robertson, "Fire and Vegetation in the Everglades," in E. V. Komarek, ed., *Tall Timbers Fire Ecology Proceedings* (Tallahassee, Fla.: Tall Timbers Research Station, 1962), No. 1, 67–80.

29. Dale L. Taylor, "Fire History and Fire Records for Everglades National Park, 1948–1979," Report T-619 (Homestead, Fla.: South Florida Research Center, 1981), 14–16.

30. Superintendent to Regional Director, June 21, 1957; Memorandum from the Director, October 9, 1957; Superintendent to Regional Director, January 6, 1958, Everglades National Park, 883.01.

31. Taylor, "Fire History and Fire Records for Everglades National Park, 1948–1979," Report T-619, 16–17.

32. E. Lowell Sumner, "The Kaweah Basin Research Reserve: An Untouched Area for the Future," Regional Director to Director, February 6, 1950, Sequoia National Park, FR 1950, 1970–1976, Ma-U, Box 327, F317, Sequoia National Park Archives.

33. Sumner, "The Kaweah Basin Research Reserve," 6; Regional Director, Memorandum of August 24, 1949, Sequoia National Park, FR 1950, 1970–1976, Ma-U, Box 327, F317, Sequoia National Park Archives.

34. Superintendent, Pipestone National Monument to Regional Director, March 21, 1950, Pipestone 701, NARA, Kansas City; Hal K. Rothman, *Managing the Sacred and the Secular: An Administrative History of Pipestone National Monument* (Omaha, Nebr.: National Park Service, 1992), 177–179.

35. Jan W. van Wagtendonk, "Dr. Biswell's Influence on the Development of Prescribed Burning in California," in *The Biswell Symposium: Fire Issues and Solutions in Urban Interface and Wildland Ecosystems* (USDA Forest Service, General Technical Report, PSW-GTR-159, 1995), 11–14; Robert Barbee, interview by Hal Rothman, part 1, November 12, 2004; Pyne, *Fire in America*, 119; David Carle, *Burning Questions: America's Fight with Nature's Fire* (Westport, Conn.: Praeger, 2002), 58–60.

36. Van Wagtendonk, "Dr. Biswell's Influence on the Development of Prescribed Burning in California," 12; Carle, *Burning Questions*, 57–58.

37. Van Wagtendonk, "Dr. Biswell's Influence on the Development of Prescribed Burning in California," 12; Carle, *Burning Questions*, 57–58.

38. Carle, *Burning Questions*, 62–63.

39. National Park Service, "Public Use of National Parks: A Statistical Report, 1941–1953"; National Park Service, "Public Use of National Parks: A Statistical Report, 1954–1964"; Charles Stevenson, "The Shocking Truth about Our National Parks," *Reader's Digest*, January 1955; Conrad L. Wirth, *Parks, Politics, and the People* (Norman: University of Oklahoma Press, 1980), 237–238.

40. Robert D. Baker, Robert S. Maxwell, Victor H. Treat, and Henry C. Dethloff, *Timeless Heritage: A History of the Forest Service in the Southwest* (Washington, D.C., 1988), 131–133; Hal K. Rothman, *Saving the Planet: The American Response to the Environment in the Twentieth Century* (Chicago: Dee, 2000), 129–130.

41. Roy E. Appleman, "A History of the National Park Service Mission 66 Program," January 1958, 1–22, NPS Technical Information Center, Denver, Colorado, microfilm.

42. Guy Fringer, *Olympic National Park: An Administrative History* (Port Angeles, Wash.: National Park Service, 1991), 102–103, 105–106, 131–132.

43. Memorandum: Forest Protection Planning, Acting Director to Regional Directors, Regions One, Two, Three, Four, and Five, July 21, 1960, Yellowstone Archives, Yellowstone Box Y-239.

44. Ibid.

45. Richard West Sellars, *Preserving Nature in the National Parks: A History* (New Haven, Conn.: Yale University Press, 1999), 212–222; A. Starker Leopold et al., "Wildlife Management in the National Parks," in James B. Trerethren, ed., *Transactions of the Twenty-Eighth North American Wildlife and Natural Resources Conference* (Washington, D.C.: Wildlife Management Institute, 1963), 1–43; National Academy of Sciences, National Research Council, "A Report by the Advisory Committee to the National Park Service on Research," typescript, August 1, 1963, http://www.cr.nps.gov/history/online_books/robbins/robbins.htm; for more on the larger changes of the era, see Patterson, *Grand Expectations*, 174–175, 277–278, 313–320; and David Halberstam, *The Fifties* (New York: Villard, 1993), 286–294, 495–496, 623–625.

46. Carle, *Burning Questions*, 43, 94, 117–119; Pyne, *Fire in America*, 293, 302.

47. Donald Swain, *Wilderness Defender: Horace M. Albright and Conservation* (Chicago: University of Chicago Press, 1970), 311–316; Richard Sellars, "The Significance of George Wright," *George Wright Forum* 17, no. 4 (2000): 46–50.

48. A. S. Leopold, S. A. Cain, C. M. Cottam, I. N. Gabrielson, and T. L. Kimball, "Wildlife Management in the National Parks," 10; Ethan Carr, *Wilderness by Design: Landscape Architecture and the National Park Service* (Lincoln: University of Nebraska Press, 1998).

49. Leopold et al., "Wildlife Management in the National Parks," 14.

50. Sellars, *Preserving Nature in the National Parks*, 214–216; Ronald A. Foresta, *America's National Parks and Their Keepers* (Washington, D.C.: Resources for the Future, 1984), 133–136, 148–162; Carr, *Wilderness by Design*, 1–14.

51. Jan van Wagtendonk, "The Evolution of National Park Service Fire Policy," 329–330; James Agee, interview by Hal Rothman, June 10, 2004.

52. Park Management Biologist to Chief Park Ranger, Memorandum: Prescribed Burn Experiment, November 18, 1965, Yellowstone National Park, N-16.

53. Ibid., 2.

54. Acting Regional Director to Superintendent, Yellowstone, December 20, 1965, N-1427 MWR, Yellowstone National Park Archives.

55. Harold H. Biswell, "Forest Fire in Perspective," in Edward V. Komarek, Sr., ed., *Tall Timbers Fire Ecology Conference Proceedings*, No. 7 (Tall Timbers, Fla.: Tall Timbers Research Station, 1967), 43–64; R. J. Hartesveldt and H. T. Harvey, "The Fire Ecology of Sequoia Regeneration," 65–78, in Edward V. Komarek, Sr., ed., *Tall Timbers Fire Ecology Conference Proceedings*, No. 7 (Tall Timbers, Fla.: Tall Timbers Research Station, 1967); Pyne, *Fire in America*, 302.

56. Sellars, *Preserving Nature in the National Parks*, 255; Art White, interview by Richard McCaslin, November 15, 1990, copy in possession of the author.

57. "Glacier National Park, Forest Fire Control Plan, March 1965," Glacier National Park, Fire Cache Records, 1910–1990, Box 2; NPS, *Fire Control Handbook*, 1965.

58. Sellars, *Preserving Nature in the National Parks*, 255–256; Kilgore interview, February 16, 2004; Agee interview, June 10, 2004.

59. Lary M. Dilsaver and William C. Tweed, *Challenge of the Big Trees: A Resource History of Sequoia and Kings Canyon National Parks* (Three Rivers, Calif.: Sequoia Natural History Association, 1990), 263–265; Sellars, *Natural Resource Management in the National Parks*, 257; Foresta, *America's National Parks and Their Keepers*, 68–74.

60. Forest Fire Control, 1966, 1–3, Yosemite National Park.

61. *Glacier Fire Film*, 1967, Glacier National Park, 1910–1984 Collection, 309-5; "Glacier National Park Forest Fire Review, November 30–December 1, 1967," Glacier National Park, 1910–1984 Collection, 309-22.

62. *Glacier Fire Film*, 3–6.

63. *Glacier Fire Film*, 8–10.

64. Pyne, *Fire in America*, 257–259.

FIVE

1. National Park Service, *Administrative Policies for Natural Areas of the National Park System* (Washington, D.C.: US Government Printing Office, 1968), 17; National Park Service, *Administrative Policies for Recreation Areas of the National Park System* (Washington, D.C.: U.S. Government Printing Office, 1968), 20–21; National Park Service, *Administrative Policies for Historical Areas of the National Park System* (Washington, D.C.: U.S. Government Printing Office, 1968), 20, 36, 60.

2. Hal K. Rothman, *The Greening of a Nation? Environmentalism in the US since 1945* (Fort Worth, Tex.: HarBrace, 1997), 161–177; David A. Adams, *Renewable Resource Policy: The Legal Institutional Foundations* (Washington, D.C.: Island, 1993), 158–163; Walter A. Rosenbaum, *The Politics of Environmental Concern*, 2d ed. (New York: Holt, Rinehart, and Winston, 1977), 48–55, 117–123; Samuel P. Hays, *Beauty, Health, and Permanence: Environmental Politics in the United States, 1955–1985* (Cambridge: Cambridge University Press, 1987), 1–25.

3. Boise Interagency Fire Agency, "The History of the Boise Interagency Fire Center," BIFC 1989, National Interagency Fire Center, miscellaneous files 1–3; Stephen J. Pyne, *Fire in America: A Cultural History of Wildland and Rural* (Princeton, N.J.: Princeton University Press, 1982), 289–294; David Carle, *Burning Questions: America's Fight with Nature's Fire* (Westport, Conn.: Praeger, 2002), 139–152; Stewart Udall, *The Quiet Crisis* (New York: Holt, Rinehart, and Winston, 1963).

4. Paul W. Hirt, *A Conspiracy of Optimism: Management of the National Forests since World War II* (Lincoln: University of Nebraska Press, 1994), 266–284; USDA Forest Service, *RARE II: A Quest for Balance in Public Lands* (Washington, D.C.: U.S. Government Printing Office, 1978); M. Rupert Cutler, "National Forests in the Balance," *American Forests* (May 1978): 1–5; M. Rupert Cutler, Introduction, *Western Wildlands: A Natural Resource Journal* 5 (Summer 1978), 1; Tim Mahoney, "RARE Draft: EIS Sparks Heavy

Input from Conservationists," *Wilderness Report* 15 (September 1978), 3–5; David Crosson, "RARE Results Final: 'An Acute Disappointment,'" *High Country News*, January 12, 1979; Richard West Sellars, *Preserving Nature in the National Parks: A History* (New Haven, Conn.: Yale University Press, 1999), 217–218; Pyne, *Fire in America*, 303; George B. Hartzog, Jr., *Battling for the National Parks* (Mt. Kisco, N.Y.: Moyer Bell Limited, 1968), 96–107; Ronald A. Foresta, *America's National Parks and Their Keepers* (Washington, D.C.: Resources for the Future, 1984), 69–70.

5. Lowell Sumner, "A History of the Office of Natural Science Studies," in *Proceedings of the Meeting of Research Scientists and Management Biologists of the National Park Service, Horace M. Albright Training Center, Grand Canyon National Park, April 6–8, 1968*, copy provided by Robert Linn, 1–6; Robert Linn, "Postscript: Current Happenings," in *Proceedings of the Meeting of Research Scientists and Management Biologists of the National Park Service, Horace M. Albright Training Center, Grand Canyon National Park, April 6–8, 1968*, copy provided by Robert Linn; James Agee interview by Hal K. Rothman, June 10, 2004.

6. Harold H. Biswell, *Prescribed Burning in California Wildlands Vegetation Management* (Berkeley: University of California Press, 1989), 58–59; Pyne, *Fire in America*, 302.

7. Bruce Kilgore interview by Hal Rothman, February 16, 2004; Jan van Wagtendonk interview by Lesley Argo, June 13, 2002.

8. Jan van Wagtendonk interview, June 13, 2002.

9. Carle, *Burning Questions*, 121–122.

10. William Briggle, historical interview by Jennifer Bottomly, September 4, 2001, Glacier National Park Archives, Oral History Collection; Harold H. Biswell, "The Role of Fire in Maintaining Forest Wilderness Quality," paper presented at the Second Annual California Plant and Soil Conference, American Society of Agronomy, California Chapter, Davis, February 1, 1973, 7.

11. Robert Barbee interview, part 1, November 12, 2004.

12. Biswell, "The Role of Fire in Maintaining Forest Wilderness Quality," 7; Biswell, *Prescribed Burning*, 109–111; James Agee interview, June 10, 2004.

13. Richard J. Hartesveldt, "Fire Ecology of the Giant Sequoia," *Natural History* (December 1964): 12–19; James Agee interview, June 10, 2004.

14. Bruce Kilgore interview, February 16, 2004; Carle, *Burning Questions*, 121, 139.

15. Carle, *Burning Questions*, 140–141; John S. McLaughlin, Memorandum from the Superintendent, May 18, 1971, Sequoia National Park Archives, Fire Records, Box 191, F1; Robert Barbee interview, part 1, November 12, 2004.

16. Bruce Kilgore interview, February 16, 2004.

17. Carle, *Burning Questions*, 140–141.

18. James Agee interview, June 10, 2004.

19. Richard Rogers Summer, "The Administration of Fire Management within the National Park Service" (M.A. thesis, California State University, Fullerton, 1978), 1–3; National Park Service, *Administrative Policies for Natural Areas of the National Park System*, 17; National Park Service, *Administrative Policies for Recreation Areas of the National Park System*, 20; National Park Service, *Administrative Policies for Historic Areas of the National Park System*, 36.

20. These questions dominated the Tall Timbers fire research conferences in the mid-1960s and early 1970s; see *Tall Timbers Fire Ecology Conference Proceedings* (Tallahassee, Fla.: Tall Timbers Research Station, 1966–1976).

21. Jan van Wagtendonk interview, June 13, 2002.

22. Bruce Kilgore interview, February 16, 2004.

23. Annual Fire Summary and Analysis, January 28, 1969, Grand Teton National Park; Annual Fire Summary and Analysis, January 14, 1970, Y-2623, Fire Records, Box 1 (located in Law Enforcement Evidence Room), Grand Teton National Park; Annual Fire Summary and Analysis, Yellowstone National Park, January 25, 1970, Yellowstone National Park Archives, Y-214.

24. "Annual Forestry Report, Yellowstone National Park, 1968," Yellowstone National Park Archives, Y-214, 1–3.

25. Ibid., "Summary 1968 Fire Season, Yosemite National Park, California," Yosemite National Park Archives, NPS Central Files, Y-14, Fire Season Summary 1968, 1–4.

26. Assistant Regional Director, Operations Western Region to Superintendents, Western Region, March 24, 1969, Subject: Forest Fires, 1968; Bruce Kilgore interview, February 16, 2004; Steve Pyne, correspondence with Hal K. Rothman, June 11, 2004.

27. Resource Management Plan, Yellowstone National Park, 1970, Box W-105, Yellowstone National Park Archives.

28. Peter H. Schuft, "A Prescribed Burning Program for Sequoia and Kings Canyon National Parks," in *Proceedings: Annual Tall Timbers Fire Ecology Conference, June 8–9, 1972* (Tallahassee, Fla.: Tall Timbers Research Station, 1972), 380–382.

29. Bruce Kilgore interview, February 16, 2004; James Agee interview, June 10, 2004.

30. Schuft, "A Prescribed Burning Program for Sequoia and Kings Canyon National Parks," 383; Larry Bancroft, Thomas Nichols, David Parsons, David Graber, Boyd Evison, and Jan [van] Wagtendonk, "Evolution of the Natural Fire Management Program at Sequoia and Kings Canyon National Parks," in *Proceedings of the Wilderness Fire Symposium, Missoula, Montana, November 15–18, 1983, Missoula, Montana.* USDA Forest Service General Technical Report INT-182, 176.

31. Schuft, "A Prescribed Burning Program for Sequoia and Kings Canyon National Parks," 380–382.

32. Ibid., 384; Acting Superintendent, Sequoia and Kings Canyon to Director, Western Region, April 10, 1970, Sequoia National Park Archives, FR 68-76, Redwood Mountain Man-U Box 34, F1; John S. McLaughlin, "Restoring Fire to the Environment in Sequoia and Kings Canyon National Parks," in *Proceedings: Annual Timbers Fire Ecology Conference, June 8–9, 1972*, 391–393.

33. Acting Superintendent, Sequoia and Kings Canyon to Director, Western Region, April 10, 1970, Sequoia National Park Archives, FR 68-76, Redwood Mountain Man-U Box 34, F1.

34. McLaughlin, "Restoring Fire to the Environment in Sequoia and Kings Canyon National Parks," 393–394.

35. Robert Barbee interview, part 1, November 12, 2004.

36. McLaughlin, "Restoring Fire to the Environment in Sequoia and Kings Canyon National Parks," 393–394; Rothman, *Saving the Planet*, 125–130.

37. Superintendent to the Park Staff, Memorandum, September 7, 1971, Sequoia National Park, Fire Records, Box 191, F1.

38. John S. McLaughlin to Richard Hartesveldt, February 10, 1970, Sequoia National Park Archives, FR 69-72, Man-U, Box 11, F92; Bancroft, Nichols, Parsons, Graber, Evison, and van Wagtendonk, "Evolution of the Natural Fire Management Program at Sequoia and Kings Canyon National Parks," 176–177.

39. James Agee interview, June 10, 2004; Carle, *Burning Questions*, 124; Biswell, *Prescribed Burning*, 112.

40. Robert Barbee interview, part 1, November 12, 2004.

41. Ibid.

42. "U.C. Foresters Aid Fire Ecology Program at Yosemite National Park," *California Agriculture* 25, no. 2 (February 1971): 3; Robert Barbee interview, part 1, November 12, 2004.

43. Harold Weaver to Lawrence C. Hadley, Superintendent, Yosemite, August 4, 1970, Yosemite National Park Archives, Central Files, Y-14, Forest Fire Control, 1970.

44. Ibid.

45. Robert Barbee, in collaboration with Harold Biswell, "Environmental Restoration Program for Yosemite National Park, 1970," Yosemite National Park Archives, NPS, Protection Division, Fire, Prescribed Fire, 4, 14, 1969–1985, 1–6.

46. James Agee interview, June 10, 2004; Rosenbaum, *Politics of Environmental Concern*, 117–123.

47. Fire Control Officer Jim Olson to Wawona District Ranger, September 22, 1970, Yosemite National Park Archives, Y-14.

48. Robert D. Barbee, "Experimental Use of Prescribed Fire to Control Lodgepole Pine Encroachment," October 9, 1970, Yosemite National Park, Y-16.

49. Carle, *Burning Questions*, 147–148; Jan van Wagtendonk interview, June 2002.

50. Jan van Wagtendonk interview, July 2002; Carle, *Burning Questions*, 150–151; Hartzog, *Battling for the National Parks*, 105.

51. *Yellowstone Fire Management Plan*, 1972; Resource Management Plan, 1975, Y-218; Robert E. Sellers and Don G. Despain, "Fire Management in Yellowstone National Park," in *Proceedings: Annual Tall Timbers Fire Ecology Conference*, vol. 14 (Tallahassee, Fla.: Tall Timbers Research Station, 1976), 102–108.

52. Jan van Wagtendonk interview, June 13, 2002; David R. Brower, *For Earth's Sake: The Life and Times of David Brower* (Salt Lake City, Utah: Gibbs Smith, 1990), 328–334.

53. No author, "Fire Management in Grand Teton National Park," Fire Records, Box 1 (located in Law Enforcement Evidence Room), Grand Teton National Park; no title, *Jackson Hole News*, June 22, 1972; no title, *Jackson Hole News*, August 10, 1972.

54. "1973 Prescribed Burning Statement," Fire Records, Box 1 (located in Law Enforcement Evidence Room), Grand Teton National Park.

55. Ibid.

56. Gary Everhardt, "To Whom It May Concern," June 28, 1974; "Testimony Presented at the Public Hearing on the Draft Environmental Impact Assessment of Grand Teton National Park's Fire-Vegetation Management Plan Held in Jackson, Wyoming on July 8, 1974"; Richard E. Baldwin to Gary Everhardt, July 10, 1974; Dale Taylor to Richard E. Baldwin, July 17, 1974; Louise G. Murie to Gary Everhardt, July 18, 1974; Philip M. Hocker to Gary Everhardt, July 19, 1974; Adolph Murie to Gary Everhardt, July 18, 1974, all in Fire Records, Box 1 (located in Law Enforcement Evidence Room), Grand Teton National Park.

57. "Narrative—Waterfalls Canyon Fire," 76; Lloyd Loope, "Report on Uh Hill Burn of 8/28/74," both in Fire Records, Box 1 (located in Law Enforcement Evidence Room), Grand Teton National Park.

58. "Has Smokey the Bear Become Smokey the Firebug?" *Jackson Hole Guide*, September 19, 1974.

59. "The Anatomy of a Public Issue: The Waterfalls Canyon Fire, Grand Teton NP" (rough draft), Fire Records, Box 1 (located in Law Enforcement Evidence Room), Grand Teton National Park.

60. "Let 'Em Burn," *Time*, October 28, 1974, p. 57.

61. Ibid.

62. National Park Service, "National Park Service Studies Show Forest Fires May Help Preserve Parks," Sequoia National Park Archives, Box 192 (Let Burn 1974).

63. Memorandum, Director to All Regional Directors and Director, National Capital Parks, March 13, 1975; Director to Harold T. Johnson, March 13, 1975; Regional Director, Western Region to Superintendents, Western Region, State Director, Hawaii, Chief, Arizona Archaeological Center, Directorate, Western Regional Office, Yosemite National Park Archives, NPS Central Files, Y-14, Forest Fire Control, 1914–1975; Foresta, *America's National Parks and Their Keepers*, 39–55.

64. Associate Director, Park System Management to Directorate, WASO Division Chiefs and All Park Superintendents, Subject: Fire Management, July 22, 1976; Interim Fire Management Program, Staff Directive 76-12, Yellowstone National Park Archives, Y-218.

65. Interim Fire Management Program, Staff Directive 76-12, July 22, 1976, Yellowstone National Park Archives, Y-218.

66. "Task Directive for Fire Management, October 1976," Yellowstone National Park Archives, Y-225; Regional Chief Scientist, Operations, Midwest Region to Research Scientist, Grand Teton National Park, December 16, 1976.

67. Associate Director, Management and Operations to Directorate and Field Directorate, Fire Management Program Responsibilities: WASO and NPS-BIFC, Staff Directive 77-1, February 24, 1977, Y-14-550, National Park Service, Harpers Ferry Center, Harpers Ferry, West Virginia.

68. Regional Director, Western Region to Superintendents, Western Region, State Director, Hawaii, Chief, Arizona Archaeological Center, Directorate, Western Regional Office, Yosemite National Park Archives, NPS Central Files, Y-14, Forest Fire Control, 1914–1975.

69. National Park Service, "Fire Management Policy, NPS-18, 1977," Glacier National Park Archives, 1910–1984 Collection, 306–8; Richard Rogers Summer, "The Administration of Fire Management within the National Park Service" (M.A. thesis, California State University, Fullerton, 1978), 13–17.

70. National Park Service, "Fire Management Policy, NPS-18, Release 1, August 23, 1979," National Park Service, Harpers Ferry Records Center, Harpers Ferry, West Virginia, 1–2.

SIX

1. Ronald Foresta, *America's National Parks and Their Keepers* (Washington, D.C.: Resources for the Future, 1984), 89–91; William Everhart, *The National Park Service* (Westport, Conn.: Greenview, 1985), 153–154.

2. USDA Forest Service, *RARE II: A Quest for Balance in Public Lands* (Washington, D.C.: U.S. Government Printing Office, 1978); M. Rupert Cutler, "National Forests in the Balance," *American Forests* (May 1978): 1–5; M. Rupert Cutler, Introduction, *Western Wildlands: A Natural Resource Journal* 5 (Summer 1978), 3; Tim Mahoney, "RARE Draft: EIS Sparks Heavy Input from Conservationists," *Wilderness Report* 15 (September 1978); David Crosson, "RARE Results Final: 'An Acute Disappointment,'" *High Country News*, January 12, 1979.

3. Environmental Assessment: Fire Management, Big Cypress National Preserve, March 1982, Technical Information Center, Denver Service Center, Denver, Colorado, D-21A, 4–8; David Carle, *Burning Questions: America's Fight with Nature's Fire* (Westport, Conn.: Praeger, 2002), 216–217.

4. James K. Agee and Harold H. Biswell, "The Fire Management Plan for Pinnacles National Monument," paper presented to the First Conference on Scientific Research in the National Parks, New Orleans, Louisiana, November 9–14, 1976, Denver Service Center, Technical Information Center, Denver, Colorado, 114/D-38, 9–13; James Agee interview, June 10, 2004.

5. Jane E. Kapler, "Glacier National Park, Assessment: Fire Management," February 1977, Denver Service Center, Technical Information Center, Denver, Colorado, 11/D-1143, sections 110–130, 271–345.2. The document is not paginated.

6. Glacier National Park, "Forest Fire Management, Glacier National Park, West Glacier," Montana, June 2, 1978, Denver Service Center, Technical Information Center, Denver, Colorado, 117/D-298, 1–3.

7. Acting Superintendent to Jim Haback, January 2, 1981; Man and the Biosphere Coordinator, Washington Office to Regional Chief Scientist, Rocky Mountain Region, March 31, 1981, Glacier National Park, Fire Collection 1910–1984, 306-11; Ron Wakimoto, "Evaluating Direct Response to Understory Burning in a Pine-Fire-Larch Forest in Glacier National Park," in Bruce Kilgore, ed., *Proceedings: National Wilderness Research Conference: Current Research*, USDA Forest Service General Technical Report INTO-212 (Missoula, Mont.: USDA Forest Service, 1986), 26–34.

8. Jan W. van Wagtendonk, "Fire Management in the Yosemite Mixed-Conifer Ecosystem," paper presented at the Symposium on Environmental Consequences of

Fire and Fuel Management in Mediterranean Ecosystems, Palo Alto, California, August 1–5, 1977, 459–463; Yosemite National Park, "Natural, Conditional, and Prescribed Fire Management Plan, 1979," Denver Service Center, Technical Service Center, Denver, Colorado, 104/D-845, 3.

9. Yosemite National Park, "Natural, Conditional, and Prescribed Fire Management Plan, 1979," 21–22.

10. Ibid., 41–43.

11. "Fire Management Plan, Sequoia and Kings Canyon National Parks, February, 1979," Denver Service Center, Technical Information Center, Denver, Colorado, 102/D-300, I-1–3, IV-5–9.

12. Superintendent, Sequoia and Kings Canyon National Parks to Regional Director, Western Region, May 24, 1979, Denver Service Center, Technical Information Center, Denver, Colorado, 102/D-330.

13. "Fire Management Plan, Sequoia and Kings Canyon National Park," February 1979, V-1–V-30.

14. Carle, *Burning Questions*, 184.

15. John Jacobs, *A Rage for Justice: The Passion and Politics of Phillip Burton* (Berkeley: University of California Press, 1995), 300–307; Barry Mackintosh, *The National Parks: Shaping the System* (Washington, D.C.: U.S. Department of the Interior, 1991), 86–89.

16. Superintendent, Custer Battlefield National Monument, "Fire Plan," July 7, 1977, Denver Service, Technical Information Center, Denver, Colorado, 381-D-31, 1–4.

17. Douglas D. Scott and Richard A. Fox, *Archeological Insights into the Custer Battlefield* (Norman: University of Oklahoma Press, 1987), xi, 108–126.

18. Memorandum, Director to Directorate and All Regional Directors, Subject: FIREPRO, NPS Normal Year Programming, December 22, 1981, Glacier National Park, Fire Collection 1910–1984, 306-11; Director to Park Managers, October 7, 1982, copy provided by Stephen J. Pyne.

19. FIREPRO, n.d., provided by Stephen J. Pyne, I-1, II-1–3.

20. Bruce Kilgore, "Introduction: Fire Management Section," in E. V. Komarek, ed., *Tall Timbers Fire Ecology Conference Proceedings* (Tallahassee, Fla.: Tall Timbers Research Station, 1976), 7–9; Hal K. Rothman, *On Rims and Ridges: The Los Alamos Area since 1880* (Lincoln: University of Nebraska Press, 1992), 273–277.

21. John Lissoway interview by Lincoln Bramwell, August 15, 2002.

22. Rothman, *On Rims and Ridges*, 275.

23. Teralene Foxx, ed., *Los Alamos Fire Symposium, Los Alamos, New Mexico, October 6–7, 1981* (Los Alamos, N.M.: Los Alamos National Laboratory, 1984), 3–6.

24. John Lissoway interview, August 15, 2002.

25. Dr. Milford R. Fletcher, conversation with the author, August 21, 1986; Senior Archaeologist Cal Cummings to Chief Anthropologist, WASO, January 24, 1986, copy in possession of the author.

26. Dr. Milford R. Fletcher, conversation with author, August 21, 1986; Diane Traylor, "Effects of La Mesa Fire on Bandelier's Cultural Resources," in Foxx, *Los Alamos Symposium*, 97.

27. Bandelier National Monument, "Interim Fire Management Plan for Bandelier National Monument," August 8, 1980, Denver Service Center, Technical Information Center, Denver, Colorado, 315/D-76, 5–6, 22–23.

28. Memorandum, Superintendent, Grand Canyon to Regional Director, Western Region, July 20, 1981, Grand Canyon National Park.

29. Rocky Mountain National Park and Shadow Mountain National Recreation Area, Wildfire Fire Management Plan Part I: Fire Control, 1–2; Rocky Mountain National Park and Shadow Mountain National Recreation Area, Wildfire Fire Management Plan Part II: Natural Fires, 1977, Denver Service Center, Technical Information Center, Denver, Colorado, D-201, 1–2.

30. Rocky Mountain National Park and Shadow Mountain National Recreation Area, Wildfire Fire Management Plan Part II: Natural Fires, 1977, 2–4; Richard D. Laven, "Natural Fire Management in Rocky Mountain National Park: A Case Study of the Ouzel Fire," in Robert M. Linn, ed., *Conference on Scientific Research in the National Parks: Proceedings of the Second Conference on Science in the National Parks, San Francisco, California, September 26–30, 1979*, 37.

31. Laven, "Natural Fire Management in Rocky Mountain National Park," 39–41.

32. Laven, "Natural Fire Management in Rocky Mountain National Park," 41; "Board of Review Report for the Ouzel Fire, Rocky Mountain National Park, Colorado, November 8–9, 1978," Glacier National Park, Fire Collection 1910–1984, 306-11.

33. "Board of Review Report for the Ouzel Fire."

34. Ibid.; Resource Conservation and Recovery Act of 1976, 42 USC 6962 (October 21, 1976).

35. "Board of Review Report for the Ouzel Fire."

36. Hal K. Rothman, *Preserving Different Pasts: The American National Monuments* (Urbana: University of Illinois Press, 1989), 226–230; Donald Craig Mitchell, *Take My Land, Take My Life: The Story of Congress's Historic Settlement of Alaska Native Land Claims, 1960–1971* (Fairbanks: University of Alaska Press, 2001), 337–542; David S. Case, *Alaska Natives and American Laws* (Fairbanks: University of Alaska Press, 1984), 14–20.

37. Frank Norris, *Alaska Subsistence: A National Park Service Management History* (Anchorage, Alaska: National Park Service, 2002), 46–161; Theodore Catton, *Land Reborn: A History of Administration and Visitor Use in Glacier Bay NP and Preserve* (Anchorage, Alaska: National Park Service, 1992), 173–190, 253–272.

38. Rothman, *Preserving Different Pasts*, 230–231; Robert Righter, *Crucible for Conservation: The Creation of Grand Teton National Park* (Niwot: Colorado Associated University Press, 1982), 103–125; William Everhart, *The National Park Service* (Boulder, Colo.: Westview, 1985).

39. Everhart, *The National Park Service*, 4; William Adams, "The Role of Fire in the Alaska Taiga: An Unsolved Problem"; "A Fresh Look at Fire," News Release, Bureau of Land Management, April 3, 1979; Pyne, *Fire in America*, 509.

40. Pyne, *Fire in America*, 303–304; William R. Moore, "From Fire Control to Fire Management," *Western Wildlands* 1, no. 3 (Summer 1974): 11–16.

41. "A Fresh Look at Fire."

42. David Kellyhouse, "Presentation to the Fire Management Subcommittee of the Alaska Land Manager's Cooperative Task Force," November 1978; William Paleck to Claire Whitlock, September 28, 1979, Alaska 6; Alaska Land Managers Cooperative Task Force, Fire Subcommittee Status Report, October 2, 1979, Alaska 15, all in National Archives, Pacific Alaska Region, RG 79, Box 1, 79-01-A1103.

43. William Paleck, Memo to Fire Organization Working Group, February 8, 1980; "Briefing Statement: Fire Management, Area of Interest: Departmental," February 19, 1981; David Butts, "Issue: NPS/BLM Interaction on Fire," all in National Archives, Pacific Alaska Region, RG 79, Box 1, 79-01-A1103.

44. "A National Park Service Career: A Conversation with John E. Cook," interview by Richard W. Sellars, January 11–12 and April 4, 2000, 63.

45. Director, Alaska to Director, Rocky Mountain Regional Office, February 13, 1980, Yellowstone National Park Archives, Y-248, Yellowstone National Park.

46. Ibid.; Pyne, *Fire in America*, 509–513.

47. "Briefing Statement: Alaska-Fire, Area of Interest: Departmental," October 19, 1981, Background; National Park Service, Alaska Region, "Issue Statement: Fire Management on National Park Service Lands in Alaska," October 22, 1981, both in National Archives, Pacific Alaska Region, RG 79, Box 1, 79-01-A1103.

48. Memorandum, Assistant Secretary for Fish and Wildlife and Parks to Assistant Secretary, Land and Water Resources, January 7, 1982, National Archives, Pacific Alaska Region, RG 79, Box 1, 79-01-A1103; Lance Gay, "Environmentalists Enlist Andrus in War against Watt," *Washington Star*, June 7, 1981, p. 3.

49. National Park Service, Alaska Region, "Issue Statement: Fire Management on National Park Service Lands in Alaska," October 22, 1981; Bureau of Land Management Press Release, "BLM Establishes Interagency Firefighting Command in Fairbanks, Alaska," January 22, 1982, National Archives, Pacific Alaska Region, RG 79, Y-14; "Interagency Fire Suppression Agreement between the Bureau of Land Management (Alaska) and the National Park Service (Alaska)," May 19, 1982, National Archives, Pacific Alaska Region, RG 79, Box 1, 79-01-A1103.

50. Memorandum, Chief, Fire Management to All Regional Directors, Attn: Fire Coordinators, April 14, 1981, Glacier National Park, Fire Collection 1910–1984, 306-11.

51. Brad Cella interview by Hal Rothman, September 27, 2002.

52. Ibid.

53. Judi Zuckert, *National Park Service: Wildland Fire Report, 1986* (Boise, Idaho: Branch of Fire Management, 1987), 1–2.

54. Ibid., 9–10.

55. Ibid., 1.

56. Judi Zuckert, *National Park Service: Wildland Fire Report, 1987* (Boise, Idaho: Branch of Fire Management, 1988), 1–2.

57. Ibid., 2.

58. Ibid.

SEVEN

1. Ronald A. Foresta, *America's National Parks and Their Keepers* (Washington, D.C.: Resources for the Future, 1984), 1–6.

2. No author, "The Yellowstone Fires: A Primer on the 1988 Fire Season, October 1, 1988," Yellowstone Y-198, 6–7; Jim Carrier, *Season of Fire* (Salt Lake City, Utah: Gibbs Smith, 1989); Rocky Barker, *Scorched Earth: How Fires in Yellowstone Changed America* (Washington, D.C.: Island, 2005), 3; David Carle, *Burning Questions: America's Fight with Nature's Fire* (Westport, Conn.: Praeger, 2002), 192; Douglas Gantenbein, *A Season of Fire: Four Months on the Firelines of the American West* (New York: Tarcher/Penguin, 2003), 128–130.

3. Robert Barbee interview by Hal Rothman, part 1, November 12, 2004.

4. Ross W. Simpson, *The Fires of 1988: Yellowstone Park and Montana in Flames* (Helena, Mont.: American Geographic Publishing, 1989), 20–22; Carrier, *Season of Fire*; Barker, *Scorched Earth*, 254–255, 277, 288.

5. "The Yellowstone Fires," 4–5; Robert Barbee interview, part 1, November 12, 2004; Stephen J. Pyne to Hal K. Rothman, e-mail, August 8, 2004.

6. "The Yellowstone Fires," 6–7; Robert Barbee interview, part 1, November 12, 2004.

7. Ibid., 6; Simpson, *The Fires of 1988*, 21–23; Robert Barbee interview, part 1, November 12, 2004.

8. "The Yellowstone Fires," 7–8.

9. Ibid., 8; Barker, *Scorched Earth*; Simpson, *The Fires of 1988*, 22–24.

10. "The Yellowstone Fires," 8; Robert Barbee interview by Hal K. Rothman, part 3, November 14, 2004.

11. Robert Barbee interview, part 3, November 14, 2004.

12. Simpson, *The Fires of 1988*, 20–21; Carrier, *Season of Fire*; Barker, *Scorched Earth*, 254, 277, 288.

13. Barker, *Scorched Earth*, 2; Simpson, *The Fires of 1988*, 25–27; Carle, *Burning Questions*, 191–195.

14. Acting Director to Directorate, Field Directorate, WASO Division Chiefs, and Park Superintendents, Subject: Fire Management Policy Review Team, July 12, 1989, Yellowstone Y-198, "Coordination and Management Review: 1988 Fires," Yellowstone National Park Archives.

15. Simpson, *The Fires of 1988*, 24; Carle, *Burning Questions*, 191; Gantenbein, *A Season of Fire*, 129–130.

16. "Yellowstone in the News: What Went Wrong in the Fires of 1988?" *Yellowstone Science* 2, no. 2 (Winter 1994): 9.

17. Ibid., 8–9; "The Yellowstone Fires," 6–7; Barker, *Scorched Earth*, 2–6.

18. Dan R. Sholly with Steve M. Newman, *Guardians of Yellowstone: An Intimate Look at the Challenges of Protecting America's Foremost Wilderness Park* (New York: Morrow, 1991), 221–222; Barker, *Scorched Earth*, 7.

19. Facts Summary of 1988, October 14, 1988, Yellowstone National Park archives, K-112.

20. Ibid.

21. Robert Barbee interview, part 3, November 14, 2004; Barker, *Scorched Earth,* 201–202.

22. Robert Barbee interview, part 3, November 14, 2004; Conrad Smith, *Media and Apocalypse: News Coverage of the Yellowstone Forest Fires, the* Exxon Valdez *Oil Spill, and the Loma Prieta Earthquake* (Westport, Conn.: Greenwood, 1992), 37–76; John Dodge, "Does National Media Coverage Represent the West Accurately?" *Olympian,* December 15, 2002; Barker, *Scorched Earth,* 213.

23. Robert Barbee interview, part 3, November 14, 2004.

24. William Penn Mott, Jr., to Sen. Malcolm Wallop, August 11, 1988, Yellowstone National Park, K-1112, L57(170), Yellowstone National Park Archives (YELL-II 57).

25. Ibid.

26. Petition to Ronald Reagan, September 7, 1988, Yellowstone K-112, Yellowstone National Park Archives (YELL-II 58).

27. David Helvarg, *The War against the Greens: The "Wise Use" Movement, the New Right, and Anti-Environmental Violence* (San Francisco: Sierra Club Books, 1994), 131; William Riebsame, Hannah Gosnell, and David Theobald, *Atlas of the New West: Portrait of a Changing Region* (New York: Norton, 1997), 103–111.

28. Barker, *Scorched Earth,* 208–209.

29. Superintendent, Yellowstone National Park to Regional Director, Rocky Mountain Region, May 8, 1989, Yellowstone Y-198, "Phase II Evaluation."

30. Ted Williams, "The Park Service and Its Burn Policy," *Rod & Reel,* March 1989, pp. 19–22.

31. Statement of William Penn Mott, Jr., Director, National Park Service, January 31, 1989, National Interagency Fire Center (hereafter NIFC), 17.

32. Thomas M. Bonnicksen, "Yellowstone Fire Information Update, Monday September 12, 1988," Yellowstone K-112 (YELL-II 59).

33. Barker, *Scorched Earth,* 202.

34. Bonnicksen, "Yellowstone Fire Information Update, Monday September 12, 1988," 2.

35. Thomas Bonnicksen, "Fire Gods and Federal Policy," *American Forests* (July–August 1989): 14–16.

36. Robert D. Barbee, Nathan L. Stephenson, David J. Parsons, and Howard T. Nichols, "Replies from the Fire Gods," *American Forests* (March–April 1990): 34–35, 70; Robert Barbee interview, part 2, November 12, 2004.

37. Stephen J. Pyne to Hal Rothman, August 8, 2004; Robert Barbee interview, part 1, November 12, 2004.

38. Robert E. Sellers and Donald G. Despain, "Fire Management in Yellowstone National Park," in *Proceedings of the Tall Timbers Fire Ecology Conference and Intermountain Fire Research Council and Land Management Symposium, Missoula, Montana* (Tallahassee, Fla.: Tall Timbers Research Station, 1976), 108; Barker, *Scorched Earth,* 228–229.

39. Robert Barbee interview, part 1, November 12, 2004; Alston Chase, *Playing God in Yellowstone: The Destruction of America's First National Park* (Boston: Atlantic Monthly Press, 1986).

40. U.S. Departments of Agriculture and the Interior, "Recommendations of the Fire Management Policy Review Team," *Federal Register* 53, no. 244 (December 20, 1988): 51196–51205; "Interagency Final Report on Fire Management Policy, May 5, 1989," in Lary Dilsaver, ed., *America's National Park System: The Critical Documents* (Lanham, Md.: Rowman and Littlefield, 1994), 418–419.

41. Department of the Interior, "New U.S. Fire Recommendations Approved by Secretaries of Interior and Agriculture," June 1, 1989.

42. Norman Christensen interview by Hal Rothman, part 2, August 17–18, 2004, 1.

43. William H. Romme and Don G. Despain, "The Yellowstone Fires," *Scientific American* 261 (November 1989): 36–44; William H. Romme, "Fire and Landscape Diversity in Subalpine Forests of Yellowstone National Park," *Ecological Monographs* 52, no. 2 (1982): 199–221; William H. Romme and Dennis H. Knight, "Landscape Diversity: The Concept Applied to Yellowstone Park," *BioScience* 32, no. 8 (1982): 664–669; William H. Romme and Don G. Despain, "Historical Perspective on the Yellowstone Fires of 1988," *BioScience* 39, no. 10 (1989): 695–699.

44. Norman Christensen interview, August 17–18, 2004, 1.

45. Ibid., 2–3; Romme and Despain, "Historical Perspective on the Yellowstone Fires of 1988," 695–699.

46. Norman Christensen interview, August 17–18, 2004, 3.

47. Ibid., 4.

48. Stephen J. Pyne to Hal K. Rothman, August 8, 2004, copy in possession of the author.

49. Ibid.

50. Ibid.

51. Stephen J. Pyne to Hal Rothman, December 18, 2004; Paul Schullery to Roberta D'Amico, October 13, 2004; Paul Schullery and Don G. Despain, "Prescribed Burning in Yellowstone National Park: A Doubtful Proposition," *Western Wildlands* 15, no. 2 (Summer 1989): 30–34.

52. Norman Christensen interview, August 17–18, 2004.

53. Sue Consolo-Murphy, "National Park Service Wildland Fire History," provided by NPS historian Janet McDonnell, December 5, 2004. Consolo-Murphy served as a resources management specialist at Yellowstone in 1988 and provided insight in her review of the manuscript.

54. Norman Christensen interview, August 17–18, 2004.

55. Robert Barbee interview, part 3, November 14, 2004.

56. Robert Barbee interview, part 2, November 12, 2004.

57. Judi Zuckert, *National Park Service Wildland Fire Report, 1989* (Boise, Idaho: Branch of Fire Management, 1990), 8; Rodney Norum, "Natural Fire Management in the National Park Service after 1988," *Renewable Resources Journal* 11, no. 1 (1993): 18.

58. Zuckert, *National Park Service Wildland Fire Report, 1989*, 11–12.

59. Statement of John M. Morehead, associate director, National Park Service, Department of the Interior, before the Subcommittee on Energy and Natural Resources, House Committee on Government Operations, on the Implementation of the Newly

Established Fire Management Policy for Federal Land Management Agencies, May 24, 1990, National Interagency Fire Center Records, Boise, Idaho.

60. Acting Director to Directorate, Field Directorate, WASO Division Chiefs, and Park Superintendents, Subject: Fire Management Policy Review Team, July 12, 1989, Yellowstone Archives.

61. Director to Directorate, Field Directorate, WASO Division Chiefs, and Park Superintendents, January 27, 1990, Revised: Wildland Fire Management, NPS-18, Prescribed Fire, January 1990, 12–40.

62. Acting NPS Fire Director, BIFC to Regional Director, Rocky Mountain Region, Attention: Fire Coordinator, May 9, 1989, Yellowstone Y-198, "Review of Fire Management Plan"; Richard H. Bahr, Assistant Fire Management Officer to Bob Barbee, Superintendent, November 13, 1991, Yellowstone Y-14, Yellowstone National Park Archives; "Yellowstone National Park Wildfire Management Plan, June 1991," Yellowstone National Park.

63. Schullery and Despain, "Prescribed Burning in Yellowstone," 30–34; John Varley, "The Status of Yellowstone's New Fire Plan," *Renewable Resources Journal* 11, no. 1 (Spring 1993): 20–21.

64. Robert Barbee interview, part 2, November 14, 2004; Lee Whittlesey to Roberta D'Amico, e-mail of November 22, 2004, provided to the author.

65. Bruce M. Kilgore, "Review Team Finds Fire Policy Sound but 'Application Needs Changing,'" Yellowstone Vertical File, n.d., ca. 1989.

66. "Statement of John M. Morehead, May 24, 1990," 3–4.

67. Ibid., 3; U.S. Departments of Agriculture and the Interior, "Recommendations of the Fire Management Policy Review Team."

68. Dean Berg, *National Park Service Wildland Fire Report, 1990* (Boise, Idaho: Branch of Fire Management, 1991), 12–13; NPS Western Region, "Yosemite Fires, 1990" (San Francisco: National Park Service, 1991), 1–7; Acting NPS Fire Director, BIFC to All National Park Service Regional and Park Fire Management Officers, June 29, 1990, Y-14 (Fire), National Interagency Fire Center, Boise, Idaho; Bruce M. Kilgore and Tom Nichols, "National Park Service Fire Policies and Programs," in James K. Brown, Robert W. Mutch, Charles W. Spoon, and Ronald H. Wakimoto, eds., *Proceedings: Symposium on Fire in Wilderness and Park Management* (Missoula, Mont.: USDA Forest Service, General Technical Report, INT-GTR-320, 1995), 24–27.

69. Kilgore and Nichols, "National Park Service Fire Policies and Programs," 24.

70. Ibid., 12–14; Dale Haggstrom, "Fire and Forest Management Policies on the Boreal Forest and Wildlife of Interior Alaska," *Wildfire* (December 1994): 31–36; Zuckert, *National Park Service Wildland Fire Report, 1989*, 11–13; Dean Berg, *National Park Service Wildland Fire Report, 1994* (Boise, Idaho: Branch of Fire Management, 1995), 43–45.

71. M. R. Dick, Jr., to Boyd Evison, February 3, 1989; Alaska Department of Natural Resources, Forestry Division, "National Fire Management Policy Review," January 1989, 3–5, National Interagency Fire Center, Yellowstone Box 2, D-131, 6639.

72. Dick to Evison, February 3, 1989.

73. NPS Fire Director BIFC to Chief, Branch of Fire and Aviation Management, July 1, 1991, National Interagency Fire Center, miscellaneous files; Ron Somerville, Deputy Commissioner, Alaska Fish and Game to Edward F. Spang, Boyd Evison, Walt Steiglitz, and Niles Caesar, April 2, 1991, NIFC miscellaneous files; Steve Holder to Russ Hansen, May 14, 1991, Evison to Somerville, May 14, 1991, draft, NIFC miscellaneous files.

74. National Park Service, *National Park Service Wildland Fire Report, 1999* (Boise, Idaho: Branch of Fire Management, 2000), 45.

75. NPS Western Region, "Yosemite Fires, 1990," 4, 17, 81.

76. Ibid., 5.

77. Ibid., 85–89; Smith, *Media and Apocalypse*, 75–77.

78. National Park Service, *National Park Service Wildland Fire Report, 1999*, 46–47, 58–60.

79. National Interagency Fire Center, *South Canyon Fire Investigation: Report of the South Canyon Fire Investigation Accident Team, August 17, 1994* (Boise, Idaho: National Interagency Fire Center, 1994), 1–4; Bret W. Butler, Roberta A. Bartlette, Larry S. Bradshaw, Jack D. Cohen, Patricia L. Andrews, Ted Putnam, and Richard J. Mangan, "Fire Behavior Associated with the 1994 South Canyon Fire on Storm King Mountain, Colorado," Research Paper RMRS-RP-9 (Ogden, Utah: U.S. Department of Agriculture, Forest Service, Rocky Mountain Research Station, 1998), 1–7; Carle, *Burning Questions*, 225; Stephen J. Pyne to Hal Rothman, e-mail, September 15, 2004.

80. Stephen J. Botti, G. Thomas Zimmerman, Howard T. Nichols, and Jan van Wagtendonk, *Fire Management and Ecosystem Health in the National Park System: Problem Analysis* (Boise, Idaho: NPS Branch of Fire and Aviation Management, 1994), 1–4.

81. "Let It Burn?" *Billings Gazette*, August 4, 1994; Don Schwennesen, "Glacier Goes for the Burn," *Missoulan*, August 7, 1994, p. B6; "GNP Managers Consider Their Options," *Glacier Reporter*, August 11, 1994; David Carkhuff, "The Howling Fire," *Hungry Horse News*, August 18, 1994,

82. "Howling Fire," *Hungry Horse News*, October 6, 1994, 5.

83. "Prescribed Natural Fire Workshop, San Francisco, California, January 3–5, 1995, Meeting Summary," NIFC, miscellaneous files.

84. Ben Jacobs, "NPS Prescribed Fire Support Modules: A Pilot Program," *Fire Management Notes* 56, no. 2 (1996): 4–5; no author, "Prescribed Fire Support Crew: Pilot Program 1995," Bandelier National Monument, Y-14 (PFSC), n.d., ca. 1995; Robert Hunter Jones, "National Park Service Prescribed Fire in the Post Yellowstone Era," *Wild Earth* (Summer 1997): 27–28.

85. U.S. Department of the Interior, U.S. Department of Agriculture, *Federal Wildland Fire Management: Policy and Program Review, Final Report, December 18, 1995* (Boise, Idaho: National Interagency Fire Center, 1995), 3–9; Carle, *Burning Questions*, 225.

86. U.S. Department of the Interior, U.S. Department of Agriculture, *Federal Wildland Fire Management: Policy and Program Review, Implementation Action Report, May 23, 1996* (Boise, Idaho: National Interagency Fire Center, 1996), iii; Resource Allocation Task Group, *Implementation of Federal Wildland Fire Management Policy: Allocation of Resources, June 15, 1998* (Boise, Idaho: National Interagency Fire Center, 1998), 4–8.

87. Carle, *Burning Questions*, 225; National Park Service, *National Park Service Wildland Fire Report, 1999*, 41.

88. National Park Service, *National Park Service 1996 Wildland Fire Report* (Boise, Idaho: National Park Service, 1997), 3–4.

EIGHT

1. Mike Davis, *Ecology of Fear: Los Angeles and the Imagination of Disaster* (New York: Metropolitan, 1998), 93–148.

2. Lincoln Bramwell, "Wilderburbs: The Rise of Rural Development in the Rocky Mountain West, 1960–2000" (Ph.D. diss., University of New Mexico, 2003); David Carle, *Burning Questions: America's Fight with Nature's Fire* (Westport, Conn.: Praeger, 2002), 225–226.

3. Steve Botti, Jim Douglas, Steve Tryson, and Wally Josephson, "Trip Report: North Rim of the Grand Canyon Fire Management Program Review, August 12–14, 1997," Grand Canyon, 1; Regional Plant/Fire Ecologist, Division of Natural Resource Management, Western Region to Regional Director, Western Region, October 23, 1981, Grand Canyon Fire Collection, Grand Canyon National Park.

4. Botti, Douglas, Tryson, and Josephson, "Trip Report," 1–2.

5. Ibid., 2–4; W. W Covington and M. M. Moore, "Post-Settlement Changes in Natural Disturbance Regimes: Implications for Restoration of Old-Growth Ponderosa Pine Ecosystems in Old-Growth Forests in the Southwest and Rocky Mountain Region," in *Proceedings of the Symposium, March 9–13, 1992, Portal, Arizona*, USDA Forest Service General Technical Report RM-213 (Fort Collins, Colo.: Rocky Mountain Forest and Range Experiment Station), 81–99; W. W. Covington and M. M. Moore, "Southwestern Ponderosa Forest Structure and Resource Conditions: Changes since Euro-American Settlement," *Journal of Forestry* 92, no. 1 (1994): 39–47; W. W. Covington and S. Sackett, "Soil Mineral Nitrogen Changes following Prescribed Burning in Ponderosa Pine," *Forest Ecology and Management* 54 (1994): 175–191; W. W. Covington, R. L. Everett, R. W. Steele, L. I. Irwin, T. A. Daer, and A. N. D. Auclair, "Historical and Anticipated Changes in Forest Ecosystems of the Inland West of the United States," *Journal of Sustainable Forestry* 4, no. 2 (1994), 13–63.

6. Botti, Douglas, Tryson, and Josephson, "Trip Report, 6–9; "Outlet Prescribed Fire April 2000," 14–23; "Fire Behavior Modeling for Outlet Prescribed Fire Project," 41–49, both in Grand Canyon, Fire Collection, Grand Canyon National Park.

7. Al Hendricks, "Outlet Fire Incident Narrative," May 2000, Grand Canyon, 1; National Park Service News Release, "Outlet Fire and Grand Canyon National Park," May 12, 2000, both in Grand Canyon Fire Collection, Grand Canyon National Park.

8. Hendricks, "Outlet Fire Incident Narrative," 1; National Park Service News Release, "Outlet Fire and Grand Canyon National Park."

9. Hendricks, "Outlet Fire Incident Narrative," 3–4; National Park Service News Release, "Outlet Fire and Grand Canyon National Park," May 13, 2000; National Park Service News Release, "Outlet Fire and Grand Canyon National Park," May 15, 2000, both in Grand Canyon Fire Collection, Grand Canyon National Park.

10. National Park Service News Release, "Suppression Continues on the Outlet Fire," May 13, 2000; National Park Service News Release, "Outlet Fire and Grand Canyon National Park," May 15, 2000; Northern Rockies Incident Management Team, "Predicted High Winds Cause Concern on the Outlet Fire," May 16, 2000; Northern Rockies Incident Management Team, "Outlet Fire Update, May 17, 2000"; Northern Rockies Incident Management Team, "Outlet Fire Update, May 18, 2000"; Northern Rockies Incident Management Team, "Outlet Fire Update, May 19, 2000"; Northern Rockies Incident Management Team, "Outlet Fire Update, May 21, 2000," all in Grand Canyon Fire Collection, Grand Canyon National Park.

11. William F. Paleck, Rodd Richardson, Bill Clark, Bill Wallis, Ron Hamilton, Stephen G. Jakla, Greg Harmon, and Tom Pittenger, "Outlet Prescribed Fire Project, Grand Canyon National Park: Investigative Report, May 2000," available at http://www.nps.gov/grca/fire/outlet/report.

12. Craig Allen, "Panel Discussion: Cerro Grande Fire," First National Congress on Fire Ecology, Prevention, and Management, San Diego, California, December 1, 2000; Craig D. Allen, R. Touchan, and Thomas Swetnam, "Landscape-Scale Fire History Studies Support Fire Management Action at Bandelier," *Park Science* 15, no. 3 (1995): 18–19.

13. Fire Investigation Team, National Interagency Fire Center, "Cerro Grande Prescribed Fire, May 4–8, 2000, Investigation Report, May 18, 2000" (Boise, Idaho: National Interagency Fire Center, 2000), 9–11.

14. Ibid., 10.

15. Ibid., 11; State of Florida, Bureau of Land Management, U.S. Forest Service, and White Mountain Apache Tribe, *Cerro Grande Prescribed Fire: Independent Review Board Report* (Washington, D.C.: National Park Service, 2000), 1–15; National Park Service, Board of Inquiry, *Cerro Grande Prescribed Fire Final Report, February 26, 2001* (Washington, D.C.: National Park Service, 2001), 42–52.

16. Fire Investigation Team, National Interagency Fire Center, "Cerro Grande Prescribed Fire, May 4–8, 2000, Investigation Report, May 18, 2000," 12.

17. Carle, *Burning Questions*, 229.

18. Fire Investigation Team, National Interagency Fire Center, "Cerro Grande Prescribed Fire," 12–13.

19. Carle, *Burning Questions*, 230; Fire Investigation Team, National Interagency Fire Center, "Cerro Grande Prescribed Fire," 13.

20. Fire Investigation Team, National Interagency Fire Center, "Cerro Grande Prescribed Fire," 13.

21. Carle, *Burning Questions*, 231–232; Gil Reavill, "Meltdown in Los Alamos," *Maxim* (October 2000).

22. Sarah Meyer, "Victims Return to Scene of Disaster," *Los Alamos Monitor*, May 16, 2000, p. 1.

23. Ibid.

24. Barry T. Hill, "Fire Management: Lessons Learned from the Cerro Grande (Los Alamos) Fire: Statement before the Committee on Energy and Natural Resources, United States Senate," July 20, 2000, 1–15, Bandelier National Monument Files, Y-14; "Report Opposes Firing Workers for Los Alamos Blaze," *USA Today*, June 20, 2001.

25. Reavill, "Meltdown in Los Alamos."

26. Board of Inquiry, *Cerro Grande Prescribed Fire Final Report,* 49.

27. Keith Easthouse, "Park Service Unfairly Scapegoated for Los Alamos Fire," *Forest Magazine* (April 5, 2001); Board of Inquiry, *Cerro Grande Prescribed Fire Final Report,* 46.

28. Board of Inquiry, *Cerro Grande Prescribed Fire Final Report,* 49.

29. Stephen J. Pyne, *Tending Fire: Coping with America's Wildfires* (Washington, D.C.: Island, 2004), 1–3.

30. NPS Wildland Fire Fact Sheet, accessed April 9, 2004, at http://www.nps.gov/fire/download/uti_abo_wildlandfirefact.pdf. The actual numbers the NPS invested were $123,741,000 in fire management and another $1,564,331,000 in ONPS funding.

31. Pyne, *Tending Fire,* 11–12.

Bibliography

Ackerman, J. "Carrying the Torch." *Nature Conservancy* 43 (1993): 16–23.

Adams, Charles. "Ecological Conditions in National Forests and in National Parks." *Scientific Monthly* 20 (June 1925): 561–590.

Agee, James K. *Fire Ecology of Pacific Northwest Forests.* Washington, D.C.: Island, 1993.

———. "Fire Management in the National Parks." *Western Wildlands* 1, no. 3 (1974): 27–33.

———. "Memorial Dedication to Dr. Harold H. Biswell." In *The Biswell Symposium: Fire Issues and Solutions in Urban Interface and Wildland Ecosystems, February 15–17, 1994.* USDA Forest Service General Technical Report PSW-GTR-158. Walnut Creek, California, Pacific Southwest Field Station, 1995, pp. 1–3.

———. "Perceptions and Professionals: Coming to Grips with Both." *Renewable Resources Journal* 11, no. 1 (Spring 1993): 25–26. "Workshop on National Parks Fire Policy: Goals, Perceptions, and Reality."

Agee, J. K., and H. H. Biswell. "Debris Accumulation in a Ponderosa Pine Forest." *California Agriculture* 24, no. 5 (1970): 6–7.

———."Seedling Survival in a Giant Sequoia Forest." *California Agriculture* 32, no. 4 (1969): 18–19.

———. "Some Effects of Thinning and Fertilization on Ponderosa Pine and Understory Vegetation." *Journal of Forestry* 68, no. 11 (1970): 709–711.

Agee, J. K., R. H. Wakimoto, and H. H. Biswell. "Fire and Fuel Dynamics of Sierra Nevada Conifers." *Forest Ecology and Management* 1 (1978): 255–265.

Ahlgren, I. F., and C. E. Ahlgren. "Ecological Effects of Forest Fires." *Botanical Review* 26 (1960): 483–533.

Allen, C. D., R. Touchan, and T. W. Swetnam. "Landscape-Scale Fire History Studies Support Fire Management Action at Bandelier." *Park Science* 15, no. 13 (1995): 18–19.

Allen, Craig D., ed. "Fire Effects in Southwestern Forests." In *Proceedings of the Second La Mesa Fire Symposium 1996.* USDA Forest Service General Technical Report RM-286.

Anderson, M. Kat. "Tending the Wilderness." *Restoration & Management Notes* 14, no. 2 (Winter 1996): 154–166.

Anderson, M. Kat, and Michael J. Moratto. "Native American Land-Use Practices and Ecological Impacts." In *Sierra Nevada Ecosystem Project Final Report to Congress*, vol. 2, *Assessments and Scientific Basis for Management Options.* Davis: University of California, Centers for Water and Wildland Resources, 1996.

Arno, Stephen F. "Eighty-eight Years of Changed in a Managed Ponderosa Pine Forest." General Technical Report RMRS-GTR-23. USDA, Forest Service, Intermountain Research Station. March 1999.

———. "Forest Fire History in the Northern Rockies." *Journal of Forestry* 78, no. 8 (August 1980): 460–465.

———. "History of Fire Occurrence in Western North America." *Renewable Resources Journal* 11, no. 1 (Spring 1993): 12–13. In special report: "National Parks Fire Policy: Goals, Perceptions and Reality."

———. "The Seminal Importance of Fire in Ecosystem Management: Impetus for This Publication." In *The Use of Fire in Forest Restoration*, General Technical Report INT-GTR-341. USDA, Forest Service, Intermountain Research Station. June 1996.

Arno, S. F., and S. Allison-Bunnell. *Flames in Our Forest: Disaster or Renewal?* Washington, D.C.: Island, 2002.

———. "Managing Fire-Prone Forests: Roots of Our Dilemma." *Fire Management Today* 63, no. 2 (2003): 12–16.

Arno, S. F., and C. C. Hardy. "The Use of Fire in Forest Restoration." In *The Use of Fire in Forest Restoration, 1996.* USDA Forest Service Intermountain Research Station General Technical Report INT-GTR-341.

Babbitt, Bruce. "Fight Fire with Fire." Address to Commonwealth Club, San Francisco, California. September 1, 1998. Available at http://www.wildfirenews.com/fire/articles/babbitt.html.

———. "Making Peace with Wildland Fire." *Wildfire* 8 (January 1999): 12–17.

Babbitt, Bruce, and Dan Glickman. "Managing the Impact of Wildfires on Communities and the Environment: A Report to the President in Response to the Wildfires of 2000." September 8, 2000. Available at http://www.whitehouse.gov/CEQ/fireport.html.

Bancroft, Larry, Thomas Nichols, David Parson, David Graber, Boyd Evison, and Jan van Wagtendonk. "Evolution of the National Fire Management Program at Sequoia and Kings Canyon National Parks." Paper presented at the Wilderness Fire Symposium, Missoula, Montana, November 15–18, 1983, pp. 174–180.

Barbee, Robert D. "Replies from the Fire Gods." *American Forests* (March–April 1990): 34–35, 70.

Barrett, Louis A. *Record of Forest and Field Fires in California.* USDA Forest Service, California Region. San Francisco, 1935.

Barro, S. C., and S. G. Conrad. "Fire Effects on California Chaparral Systems: An Overview." *Environment International* 76 (1991): 135–149.

Barry, W. James, and R. Wayne Harrison. "Prescribed Burning in the California State Park System." Paper presented at the Symposium on Fire in California Ecosystems:

Integrating Ecology, Prevention, and Management, San Diego, California, November 17–20, 1997.

Baskin, Yvonne. "Yellowstone Fires: A Decade Later; Ecological Lessons Learned in the Wake of the Conflagration." *BioScience* 49, no. 2 (February 1999).

Baumgartner, David M., et al., eds. *Prescribed Fire in the Intermountain Region.* Pullman: Washington State University, 1989.

Biswell, Harold. "The Big Trees and Fire." *National Parks Magazine*, April 11–14, 1961.

———. "Danger of Wildfires Reduced in Ponderosa Pine." *California Agriculture* 4, no. 10 (1960): 5–6.

———. "Fire Ecology in Ponderosa Pine–Grassland." In *Proceedings: Annual Tall Timbers Fire Ecology Conference, June 8–9, 1971*, pp. 69–96. Tallahassee, Fla.: Tall Timbers Research Station, 1972.

———. "Fire Ecology: Past, Present, and Future." Keynote talk to the Ecology Section, American Association for the Advancement of Science, Davis, California, June 23, 1980.

———. "Forest Fire in Perspective." In *Proceedings from the California Tall Timbers Fire Ecology Conference, November 9–10, 1967*, pp. 43–63.

———. "Litter Production." *California Agriculture* 20 (1966): 5–7.

———. "Man and Fire in Ponderosa Pine." *Sierra Club Bulletin* 44, no. 7 (1959): 44–53.

———. *Prescribed Burning in California Wildlands Vegetation Management.* Berkeley: University of California Press, 1989.

———. "Prescribed Burning in Georgia and California Compared." *Journal of Range Management* 11, no. 6 (1958): 293–298.

———. "Prescribed Fire as a Management Tool." Paper presented at the Symposium on Environmental Consequences of Fire and Fuel Management in Mediterranean Ecosystems, Palo Alto, California, August 1–5, 1977.

———. "Reduction of Wildfire Hazard." *California Agriculture* 13, no. 6 (1959): 5.

———. "Research in Wildland Fire Ecology in California." In *Proceedings: First Tall Timbers Fire Ecology Conference, March 1–2, 1962*, pp. 63–97. Tallahassee, Fla.: Tall Timbers Research Station, 1963.

———. "The Role of Fire in Maintaining Forest Wilderness Quality." Paper presented at the Second Annual California Plant and Soil Conference, California Chapter, American Society of Agronomy, February 1, 1973.

———. "The Use of Fire in Wildland Management." In S. V. Wangtrup and James J. Parsons, eds., *Natural Resources: Quality and Quantity.* Berkeley: University of California Press, 1967.

———. "Some Aspects of Simulated Natural Fires in Vegetation Management." Society for Range Management. Omaha, Nebraska February 19, 1976. Session introductory comments.

Biswell, Harold H., Harry R. Kallander, Roy Komarek, Richard J. Vogl, and Harold Weaver. "Ponderosa Fire Management: Miscellaneous Publication No. 2." Tallahassee, Fla.: Tall Timbers Research Station, 1973.

Biswell, Harold, and Harold Weaver. "Redwood Mountain." *American Forests* 74 (1968): 20–23.

Blumer, J. C. "Fire as a Biological Factor." *Plant World* 13 (1910): 42–44.

Bock, J. H., and C. E. Bock. "Natural Reforestation in the Sierra Nevada-Donner Ridge Burn." In *Proceedings from the Annual Tall Timbers Fire Ecology Conference, April 10–11, 1969*, pp. 119–126.

Boerker, Richard H. "Light Burning versus Forest Management in Northern California." *Journal of Forestry* 10 (1912): 184–194.

Bolgiano, Chris. "Yellowstone and the Let-Burn Policy." *American Forests* (January–February 1989): 22–25, 74–78.

Bond, W. J., and B. W. van Wilgen. *Fire and Plants* (London: Chapman & Hall, 1996).

Bonnicksen, Thomas M. *America's Ancient Forest: From the Ice Age to the Age of Discovery.* New York: Wiley, 2000.

———. "Fire Gods and Federal Policy." *American Forests* (July–August 1989): 14–16, 66–68.

Botti, S. J. "Funding Fuels Management in the National Park Service: Costs and Benefits." In D. R. Weise and R. E. Martin, eds., *Biswell Symposium: Fire Issues and Solutions in Urban Interface and Wildland Ecosystems, 15–17 February 1994, Walnut Creek, California.* USDA Forest Service General Technical Report
PSW-GTR-158.

———. "The National Park Service Wildland Fire Management Program." In A. Gonzalez-Caban and P. N. Omi, eds., *Proceedings of the Symposium on Fire Economics, Planning, and Policy: Bottom Lines, April 5–9, 1999, San Diego, California.* USDA Forest Service Pacific Southwest Research Station PSW-GTR-173.

Botti, S. J., and H. T. Nichols. "Availability of Fire Resources and Funding for Prescribed Natural Fire Programs." In J. K. Brown, R. W. Mutch, C. W. Spoon, and R. H. Wakimoto, eds., *Proceedings: Symposium on Fire in Wilderness and Park Management, 30 March–1 April 1993, Missoula, Montana.* USDA Forest Service General Technical Report INT-GTR-320.

Botti, S. J., P. N. Omi, and D. B. Rideout. "An Analytical Approach for Assessing Cost Effectiveness of Landscape Prescribed Fires." In A. Gonzalez-Caban and P. N. Omi, eds., *Proceedings of the Symposium on Fire Economics, Planning, and Policy: Bottom Lines, April 5–9, 1999, San Diego, California.* USDA Forest Service Pacific Southwest Research Station PSW-GTR-173.

Boyd, Robert. *Indians, Fire and the Land in the Pacific Northwest.* Corvallis: Oregon State University Press, 1999.

Bradley, C. B. "Some Problems Relating to Big Trees." *American Forests* 77 (1971): 29–31.

Brennan, Leonard A., and Sharon M. Hermann. "Prescribed Fire and Forest Pests: Solutions for Today and Tomorrow." *Journal of Forestry* (November 1994): 34–36.

Brown, Arthur A., and Kenneth P. Davis. *Forest Fire: Control and Use*, 2d ed. New York: McGraw-Hill, 1973.

Brown, James K., et al., eds. *Proceedings: Fire in Wilderness and Parks Symposium, 1995.* USDA Forest Service General Technical Report INT-GTR-320.

Bruce, Donald. "Light-Burning: Report of the California Forestry Committee." *Journal of Forestry* 21 (1928): 129–133.

Butler, Mary Ellen. *Prophet of the Parks: The Story of William Penn Mott, Jr.* Ashburn, Va.: National Recreation and Park Association, 1999.

Cammerer, Arno B. "Outdoor Recreation: Gone with the Flames." *American Forests* (April 1939): 182.

Caprio, Anthony C., and David M. Graber. "Returning Fire to the Mountains: Can We Successfully Restore the Ecological Role of Pre-Euroamerican Fire Regimes to the Sierra Nevada?" In David N. Cole and Stephen F. McCool, eds., *Proceedings: Wilderness Science in a Time of Change.* Proc. RMRS-P-000. USDA, Forest Service, Rocky Mountain Research Station. Ogden, Utah, 2000.

Carle, David. *Burning Questions: America's Fight with Nature's Fire.* Westport, Conn.: Praeger, 2002.

Cermak, Robert W. "Fire Control in the National Forests of California, 1898–1920." Master's thesis, California State University, Chico, 1986.

"Cerro Grande . . . Facing the Flames." Cerro Grande Fire Special Edition. *Los Alamos Monitor*, June 18, 2000.

Chandler, Craig, et al. *Fire in Forestry*, 2 vols. New York: Wiley, 1983.

Chapman, H. H. "Editorials: Fire: Master or Servant." *Journal of Forestry* 37 (1931): 605.

———. "Fires and Pines: A Realistic Appraisal of the Role of Fire in Reproducing and Growing Southern Pines." *American Forests* 50 (1944): 62–64.

———. "Forest Fires and Forestry in the Southern States." *American Forests* 18 (1912): 510–517.

———. "Prescribed Burning in the Loblolly Pine Type." Exchange of letters with William L. Hall. *Journal of Forestry* 45 (1947): 209–212.

———. "Prescribed Burning versus Public Forest Fire Services." *Journal of Forestry* 45 (1947): 804–808.

———. "Some Further Relations of Fire to Longleaf Pine." *Journal of Forestry* 30 (1932): 602–604.

———. "To Whom It May Concern." *American Forests* 62, no. 8 (August 1956): 54–55.

Chapman, H. H., and H. N. Wheeler. "Controlled Burning." *Journal of Forestry* 39 (1941): 886–891.

Christensen, N. L., L. Cotton, T. Harvey, R. Martin, J. McBride, P. Rundel, and R. Wakimoto. "Review of Fire Management Program for Sequoia–Mixed Conifer Forests of Yosemite, Sequoia and Kings Canyon National Parks." Final report to National Park Service. Washington, D.C., 1987.

Christensen, Norman L., James K. Agee, Peter F. Brussard, Jay Hughes, Dennis H. Knight, G. Wayne Minshall, James M. Peek, Stephen J. Pyne, Frederick J. Swanson, Jack Ward Thomas, Stephen Wells, Stephen E. Williams, and Henry A. Wright. "Interpreting the Yellowstone Fires of 1988." *BioScience* 39, no. 10 (November 1989): 678–685.

Clairborne, R. "Can There Be a 'Good' Forest Fire?" *Smithsonian* (May 1972).

Clar, C. Raymond. *California Government and Forestry: From Spanish Days until the Creation of the Department of Natural Resources in 1927.* Sacramento: California State Board of Forestry, 1959.

———. "The Development of a Forest Fire Protection System in the California Division of Forestry, 1930–42." An interview by Mrs. Amelia Fry. Regional Oral History Office, Berkeley, University of California, May 29, 1966.

Clark, J. S. "The Forest Is for Burning." *Natural History* (January 1989): 51–53.

Cohen, Jack D. "Reducing the Wildland Fire Threat to Homes: Where and How Much?" In *Symposium on Fire Economics, Policy, and Planning: Bottom Lines, April 5–9 1999, San Diego, California.* USDA Forest Service General Technical Report PSW-GTR-173.

———. "Why Los Alamos Burned." *Forest Magazine* (2000).

Coman, Warren E. "Did the Indian Protect the Forest?" *Pacific Monthly* 26, no. 3 (September 1911): 300–309.

Conarro, R. M. "Fighting Tomorrow's Fires Today." *American Forests* (April 1939): 214.

Cooper, C. F. "Changes in Vegetation, Structure, and Growth of Southwestern Pine Forest since White Settlement." *Ecological Monographs* 30, no. 3 (Summer 1960): 129–164.

———. "The Ecology of Fire." *Scientific American* (April 1961).

Covington, W. Wallace. "Ponderosa Ecosystem Restoration Institute, Northern Arizona University School of Forestry, 1999. Available at http://www.eri.nau.edu/cov99vfn.html

Cowles, Raymond B. "Fire Suppression, Faunal Changes and Condor Diets." In *Proceedings: Tall Timbers Fire Ecology, No. 7*, pp. 217–224. Tallahassee, Fla.: Tall Timbers Research Station, 1967.

———. "Starving the Condor." *California Fish and Game* 44 (1958): 175–181.

Crosby, Bill. "Our Wild Fire: History Shows That Nearly All of California Is Designed to Burn." *Sunset* (June 1992): 64–72.

Crutzen, P. G., and J. G. Goldammer, eds. *Fire in the Environment: Its Ecological, Climatic, and Atmospheric Chemical Importance.* New York: Wiley, 1993.

Dana, Samuel T. *Forest and Range Policy: Its Development in the United States.* New York: McGraw Hill, 1956.

Daniels, Orville L. "Fire Management Takes Commitment." In *Proceedings: Tall Timbers Fire Ecology Conference and Fire and Land Management Symposium, October 8–10, 1974.* Tallahassee, Fla.: Tall Timbers Research Station, 1976.

———. "A Forest Supervisor's Perspective on the Prescribed Natural Fire Program." In *Proceedings: 17th Tall Timbers Fire Ecology Conference, May 18–21, 1989. High Intensity Fire in Wildlands.* Tallahassee, Fla.: Tall Timbers Research Station, 1989.

———. "Test of a New Land Management Concept: Fritz Creek 1973." *Western Wildlands* 1, no. 3 (1974): 23–26.

Davis, James B., and Robert E. Martin, eds. *Proceedings of the Symposium on Wildland Fire 2000.* USDA Forest Service General Technical Report PSW-101.

Davis, Tony. "The West's Hottest Question: How to Burn What's Bound to Burn." *High Country News* 32, no. 11 (June 5, 2000): 3.

Dawson, Kerry J., and Steven E. Grego. "The Visual Ecology of Prescribed Fire in Sequoia National Park." In *Proceedings of the Symposium on Giant Sequoias: Their*

Place in the Ecosystem and Society, June 23–25, 1992, Visalia, California. USDA Forest Service Gel Tech Rep. PSW-151: 99–107, 1994.

DeBuys, William. "Los Alamos Fire Offers a Lesson in Humility." *High Country News* 32, no. 13 (July 3, 2000): 7.

"A Defense of Forest Fires." *Literary Digest,* August 9, 1913.

Demmon, E. L. "Fires and Forest Growth." *American Forests* 35 (April 1929).

Despain, D. *Yellowstone Vegetation: Consequences of Environment and History in a Natural Setting.* Boulder, Co: Roberts Rinehart, 1990.

Despain, D., ed. "Plants and Their Environments." *Proceedings of the First Scientific Conference on the Greater Yellowstone Ecosystem.* Technical Report NPS/NRYELL/NRTR, 1994.

Despain, D., A. Rodman, P. Schullery, and H. Schovic. "Burned Area Survey of Yellowstone National Park." In *The Fires of 1988.* Division of Research and Geographic Information Systems Laboratory, Yellowstone National Park, 1989.

Despain, D., and W. H. Romme. "Historical Perspective on the Yellowstone Fires of 1988." *BioScience* 32 (1982): 695–699.

Despain, Don G., and William H. Romme. "Ecology and Management of High-Intensity Fires in Yellowstone National Park." In *Proceedings: 17th Tall Timbers Fire Ecology Conference, May 18–21, 1989: High Intensity Fire in Wildlands,* pp. 43–57. Tallahassee, Fla.: Tall Timbers Research Station.

Devlin, Sherry. "Check with Reality: Intense Blazes of 2000 May Be Wake-up Call to Return Fire to Forests." *Missoulan.* August 22, 2000, Missoula, Montana. Available at http://www.fs.fed.us/rm/main/pa/newsclips/00_08/082200_reality.html.

Doxey, Wall. "Fire or Forestry: The South's Great Problem." *American Forests* (April 1939): 161.

duBois, Coert. "Cooperative Brush-Burning in the California National Forests." Draft circular, 95–97–03, Box 23 (27837), "Fire, Coop. 1915–23." NARA, San Bruno, California, 1915.

———. *Systematic Fire Protection in the California Forests.* USDA, Forest Service, Washington, D.C.: Government Printing Office, 1914.

Easthouse, Keith. "Los Alamos Inferno: Los Alamos National Laboratory Has Long Played with Nuclear Fire, but Can It Handle a Forest Fire?" *Forest Magazine* (September–October 1999). Available at http://www.forestmag.org/losalamosfire.htm.

Elfring, Chris. "Yellowstone: Fire Storm over Fire Management." *BioScience* 39, no. 10 (November 1989): 667–672.

Evans, C. F. "Can the South Conquer the Fire Scourge?" *American Forestry* 50 (May 1944): 227–229.

"The Father of Smokey Bear Speaks." *Forest Log* (September–October 1994): 14–17.

"The Fire Next Time." *Time,* August 7, 1972, pp. 48–49.

Flader, S. L., and J. B. Callicott, eds. *The River of the Mother of God and Other Essays by Aldo Leopold.* Madison: University of Wisconsin Press, 1991.

Floyd, Donald W. *Forest Sustainability: The History, the Challenge, the Promise.* Durham, N.C.: Forest History Society, 2002.

Folweiler, A. D. "The Place of Fire in Southern Silviculture." *Journal of Forestry* 50 (1952): 187–190.

Franke, Mary Ann. *Yellowstone in the Afterglow: Lessons from the Fires.* Mammoth Hot Springs, Wyo.: Yellowstone Center for Resources, Yellowstone National Park, 2000.

Gabrielson, Ria N. "Burning Wildlife." *American Forests* (April 1939): 186.

Gannett, Henry, ed. "Report of the National Conservation Commission, February, 1909; Special Message from the President of the United States." Washington, D.C.: Government Printing Office, 1909.

Gillette, Charles A. "Campaigning against Forest Fires." *American Forests* (April 1931): 209, 256.

Goudsblom, Johan. *Fire and Civilization.* New York: Penguin, 1992.

Graber, D. M. "Coevolution of National Park Service Fire Policy and the Role of National Parks." In J. F. Lotan, B. M. Kilgore, W. C. Fischer, and R. F. Mutch, eds., *Proceedings: Symposium and Workshop on Wilderness Fire, 15–18 November 1983, Missoula, Montana.* USDA Forest Service General Technical Report INT-182.

Graber, D. M., and D. P. Parsons. "Twenty-six Years of Prescribed Fire Management in Sequoia and Kings Canyon National Parks: What Has Been Accomplished in Restoring Fire and Its Effects?" In Leonard A. Brennan and Teresa L. Pruden, eds., *Fire in Ecosystem Management: Shifting the Paradigm from Suppression to Prescription: Proceedings of the Tall Timbers Fire Ecology Conference, No. 20.* Tallahassee, Fla.: Tall Timbers Research Station, 1998.

Graves, Henry S. "D-5: Fire Cooperation Brush Burning." Letter to Coert duBois, 95-97-03, Box 23, "Fire, Coop. 1915–1923." National Archives and Records Administration, San Bruno, California, January 24, 1918.

———. "The Forest Service and Light-Burning Experiments." *American Lumberman* 32, no. 2 (February 1928): 76–77.

———. "Graves Terms Light Burning 'Paiute Forestry'." *Timberman* (January 1920): 35.

———. "National Forests and National Parks in Wildlife Conservation." Proceedings of the National Parks Conference. Washington, D.C.: Government Printing Office, 1917.

———. *Protection of Forests from Fire.* U.S. Department of Agriculture, Forest Service, Bulletin No. 82. Washington, D.C.: Government Printing Office, 1910.

———. "The Torch in the Timber: It May Save the Lumberman's Property, but It Destroys the Forests of the Future." *Sunset: The Pacific Monthly* 44 (April 1920): 37–40, 80–90.

Graves, Walter L., and Gary Reece. "The Legacy of Harold Biswell in Southern California: His Teaching Influence on the Use of Prescribed Fire." USDA Forest Service General Technical Report PSW-GTR-158, 1995. First presented at the Biswell Symposium: Fire Issues and Solution in Urban Interface and Wildland Ecosystems, February 15–27, 1994, Walnut Creek, California.

Greeley, W. B. "'Piute Forestry' or the Fallacy of Light Burning." *Timberman* (March 1920): 38–39.

Green, Lisle R. *Burning by Prescription in Chaparral.* PSW-51. USDA-Pacific Southwest Forest and Range Experiment Station, Berkeley, California, May 1981.

Green, S. W. "The Forests That Fire Made." *American Forests* 37 (1931): 583–584, 618.

Gruell, George E. *Fire in Sierra Nevada Forests: A Photographic Interpretation of Ecological Change since 1849.* Missoula, Mont.: Mountain, 2001.

———. "Indian Fires in the Interior West: A Widespread Influence." Paper presented at the Wilderness Fire Symposium, Missoula, Montana, November 15–18, 1983.

———. "A Prerequisite for Better Public Understanding of Fire Management Challenges." In *Proceedings: 17th Tall Timbers Conference, May 21, 1989: High Intensity Fire in Wildlands,* pp. 25–38. Tallahassee, Fla.: Tall Timbers Research Station, 1989.

Guth, A. Richard, and Stan B. Cohen. *Red Skies of '88.* Missoula, Mont.: Pictorial Histories, 1989.

Haggerty, P. K. "Fire Effects in Blue Oak Woodland." In *Proceedings of the Symposium on Oak Woodlands and Hardwood Rangeland Management, 1991.* USDA Forest Service PSWFRES General Technical Report PSW-126.

Halvorson, William L., and Gary E. Davis. *Science and Ecosystem Management in the National Parks.* Tucson: University of Arizona Press, 1996.

Hampton, H. Duane. *How the U.S. Cavalry Saved Our National Parks.* Bloomington: Indiana University Press, 1971.

Hanson, Chad. "The Big Lie: Logging and Forest Fires." *Earth Island Journal* 15, no 1 (Spring 2000).

Harper, Roland M. "A Defense of Forest Fires." *Literary Digest* (August 1913): 9.

Harrison, Robert Pogue. *Forests: The Shadow of Civilization.* Chicago: University of Chicago Press, 1992.

Hartesveldt, R. J. "Fire Ecology of the Giant Sequoia: Controlled Fires May Be One Solution to the Survival of the Species." *Natural History Magazine* 73 (1964): 12–19.

Hartesveldt, R. J., and H. T. Harvey. "The Fire Ecology of Sequoia Regeneration." In *Proceedings from the Tall Timbers Fire Ecology Conference, November 9–10, 1967,* pp. 64–78.

———. "Sequoia's Dependence on Fire." *Science* 166 (1969): 552–553.

Hays, Samuel P. *Beauty, Health, and Permanence: Environmental Politics in the United States, 1955–1985.* Cambridge: Cambridge University Press, 1987.

———. *Conservation and the Gospel of Efficiency: The Progressive Conservation Movement, 1890–1920.* Cambridge, Mass.: Harvard University Press, 1959.

———. *A History of Environmental Politics since 1945.* Pittsburgh, Pa.: University of Pittsburgh Press, 2000.

Hazen, Margaret Hindle, and Robert M. Hazen. *Keepers of the Flame: The Role of Fire in American Culture 1775–1925.* Princeton, N.J.: Princeton University Press, 1992.

Heady, H. F. "Burning and the Grasslands in California." In *Proceedings of the Tall Timbers Fire Ecology Conference, 1973,* pp. 97–107.

Heinrichs, Jay. "The Ursine Gladhander." *Journal of Forestry* (October 1982): 642.

Heinselman, M. L. "Preserving Nature in Forested Wilderness Areas and National Parks." *National Parks & Conservation Magazine* 44 (1970): 8–14.

Hensel, R. I. "Recent Studies on the Effect of Burning on Grassland Vegetation." *Ecology* 4 (1923): 183–188.

Hester, Eugene. "The Evolution of Park Service Fire Policy." *Renewable Resources Journal* 11, no. 1 (Spring 1993): 14–15. In special report: "Workshop on National Parks Fire Policy: Goals, Perceptions, and Reality."

Heyward, Frank, Jr. "History of Forest Fires in the South." *Forest Farmer* 9, no. 8 (1950): 3, 10–11.

Hill, Barry T. *Fire Management: Lessons Learned from the Cerro Grande (Los Alamos) Fire.* GAO/T-RCED-00-257. Washington, D.C.: General Accounting Office, 2000.

———. *Western National Forests: A Cohesive Strategy Is Needed to Address Catastrophic Wildfire Threats.* GAO-RCED-99-65. Washington, D.C.: General Accounting Office, 1999.

Holbrook, Stewart H. B*urning an Empire: The Story of American Forest Fires.* New York: Macmillan, 1942.

Hoxie, George L. "How Fire Helps Forestry: The Practical vs. the Federal Government's Theoretical Ideas." *Sunset* 34 (August 1910): 145–151.

Huggard, Christopher J., and Arthur R. Gomez, eds. *Forests under Fire: A Century of Ecosystem Mismanagement in the Southwest.* Tucson: University of Arizona Press, 2001.

Hurley, Harry. "Prescribed Burning in the 21st Century." USDA Forest Service General Technical Report PSW-GTR-158, 1995. Abbreviated version presented at the Biswell Symposium: Fire Issues and Solutions in Urban Interface and Wildland Ecosystems, Walnut Creek, California, February 15–17, 1994.

International Association of Wildland Fire. *Bibliography of Wildland Fire.* Hot Springs, S.D.: IAWF, 1996.

Jenks, Cameron. *The Development of Government Forest Control in the United States.* Baltimore, Md.: Johns Hopkins University Press, 1928.

Jepson, W. L. "The Fire-Type Forest of the Sierra Nevada." *Intercollegiate Forestry Club Annual* 1, no. 1 (1921): 7–10.

Johnson, K. Norman, John Sessions, Jerry Franklin, and John Gabriel. "Integrating Wildfire into Strategic Planning in Sierra Nevada Forests." *Journal of Forestry* (January 1998): 42–49.

Johnson, Von J. "Prescribed Burning: Requiem or Renaissance?" *Journal of Forestry* 82, no. 2 (1984): 82–90.

Kay, Charles E. "Aboriginal Overkill and Native Burning: Implications for Modern Ecosystem Management." Available at http://wings.buffalo.edu/academic/department/anthropolgy/documents/burning.

Keeley, J. "Fire and Invasive Plants in California Ecosystems." *Fire Management Today* 63 (1999): 18–19.

Keifer, Mary Beth. "Fuel Load and Tree Density Changes following Prescribed Fire in the Giant Sequoia–Mixed Conifer Forest: The First 14 Years of Fire Effects Monitoring." In *Tall Timbers Fire Ecology Conference: 20th Proceedings: Fire in Ecosystem Management: Shifting the Paradigm from Suppression to Prescription*, pp. 306–309. Tallahassee, Fla.: Tall Timbers Research Station, 1998.

Kennedy, Roger. "Fires Illuminate Our Illusions in the Southwest." *High Country News* 32, no. 13 (July 3, 2000): 7.

Kerscher, J. R., and M. C. Axelrod. "A Process Model of Fire Ecology and Succession in a Mixed-Conifer Forest." *Ecology* 65 (1984): 1735–1742.

Kilgore, Bruce. "The Ecological Role of Fire in Sierra Conifer Forests; Its Application to National Park Management." *Quaternary Research* 3 (1973): 496–513.

———. "Fire in Ecosystem Distribution and Structure: Western Forests and Shrublands." In *Fire Regimes and Ecosystem Properties 1981*. USDA Forest Service General Technical Report GTR-WO-26.

———. "Fire Management in the National Parks: An Overview." In *Tall Timbers Fire Ecology Conference, October 8–10, 1976*, pp. 45–57.

———. "From Fire Control to Fire Management: An Ecological Basis for Policies." *Transactions: North American Wildlife and Natural Conferences* 41 (1976): 477–483.

———. "Integrated Fire Management in National Parks." *Proceedings of 1975 National Convention, Society of American Foresters* (1976): 178–188.

———. "Introduction: Fire Management Section." *Tall Timbers Fire Ecology Conference, October 8–10, 1976*, pp. 7–9.

———. "Research Needed for an Action Program of Restoring Fire to Giant Sequoias." *Intermountain Fire Research Council Symposium on The Role of Fire in the Intermountain West* (1970): 172–180.

———. "Restoring Fire to High Elevation Forests in California." *Journal of Forestry* 70 (1972): 266–271.

———. "Restoring Fire to National Park Wilderness." *American Forests* (1985): 16–19, 57–59.

———. "Restoring Fire to the Sequoias." *National Parks & Conservation Magazine* 44 (1970): 16–22.

———. "The Role of Fire in a Giant Sequoia–Mixed Conifer Forest." *Research in Parks* (1971): 93–116.

Kilgore, Bruce, and H. H. Biswell. "Seedling Germination following Fire in a Giant Sequoia Forest." *California Agriculture* 25 (February 1971): 8–10.

Kitts, Joseph A. "Forest Destruction Prevented by Control of Surface Fires." *American Forestry* 25 (1919): 1284–1306.

———. "Preventing Forest Fires by Burning Litter." *Timberman* (July 1919): 91.

Knight, D. H. "Parasites, Lightning, and the Vegetation Mosaic in Wilderness Landscapes." In M. G. Turner, ed., *Landscape Heterogeneity and Disturbance*. New York: Springer-Verlag, 1987.

Knight, D. H., and W. H. Romme. "Landscape Diversity: The Concept Applied to Yellowstone Park." *BioScience* 32 (1982): 664–670.

Knight, D. H., and L. Wallace. *Mountains and Plains: The Ecology of Wyoming Landscapes*. New Haven, Conn.: Yale University Press, 1994.

———. "The Yellowstone Fire Controversy." In R. Keiter and M. Boyce, eds., *The Greater Yellowstone Ecosystem: Redefining America's Wilderness Heritage*. New Haven, Conn.: Yale University Press, 1991.

———. "The Yellowstone Fires: Issues in Landscape Ecology." *BioScience* 39 (1989): 700–706.

Knudsen, T. "Feeding the Flame." *Sacramento Bee,* Special Report, November 27–December 1, 1994.

Koch, Elers. "The Passing of the Lolo Trail." *Journal of Forestry* 33 (1935): 98–104.

Koehler, Matthew. "The Truth about Logging and Wildfire Prevention." *Wild Rockies Organization.* Available at http://www.wildrockies.org/wildfire.

Komarek, E. V. "Comments on the History of Controlled Burning in the Southern United States." In *Proceedings: 17th Annual Arizona Watershed Symposium.* Arizona Water Commission Report No. 5, Phoenix, Arizona, September 19, 1973.

———. "A Quest for Ecological Understanding: The Secretary's Review, March 15, 1958–June 30, 1975." Miscellaneous Publications No. 5. Tallahassee, Fla.: Tall Timbers Research Station, 1977.

———. "Reflections by E. V. Komarek, Sr." In *Proceedings: 17th Tall Timbers Fire Ecology Conference, May 18–21, 1989: High Intensity Fire in Wildland Management, Challenges and Options.* Tallahassee, Fla.: Tall Timbers Research, 1991.

———. "The Use of Fire: An Historical Background." In *Proceedings: First Annual Tall Timbers Fire Ecology Conference.* Tallahassee, Fla.: Tall Timbers Research Institute, 1962.

Kotok, E. I. "Fire: A Major Ecological Factor in the Pine Region of California." *Proceedings of the Fifth Pacific Science Congress, Canada.* Toronto: University of Toronto Press, 1934.

———. "Fire: A Problem in American Forestry." *Scientific Monthly* 31 (1930): 450–452.

Kunzig, R. "These Woods Are Made for Burning." *Discover* (1988): 86–95.

Lambert S., and T. J. Stolingren. "Giant Sequoia Mortality in Burned and Unburned Stands: Does Prescribed Burning Significantly Affect Mortality Rates?" *Journal of Forestry* 86 (1988): 44–46.

Landers, J. Larry. "About E. V. Komarek, Sr." In *Proceedings: 17th Tall Timbers Fire Ecology Conference, May 18–21, 1989: High Intensity Fire in Wildland Management,* pp. 3, 4. Tallahassee, Fla.: Tall Timbers Research, 1991.

Langston, Nancy. *Forest Dreams, Forest Nightmares: The Paradox of Old Growth in the Inland West.* Seattle: University of Washington Press, 1995.

Larson, G. B. "Whitaker's Forest." *American Forests* 72, no. 9 (1966): 22–25, 40–42.

Lawrence, G. E. "Ecology of Vertebrate Animals in Relation to Chaparral Fire in the Sierra Nevada Foothills." *Natural Resources: Quality and Quantity* 47 (1966): 278–291.

Lawrence, G., and H. Biswell. "Effect of Forest Manipulation on Deer Habitat in Giant Sequoia." *Journal of Wildlife Management* 36 (1972): 595–605.

Lee, Robert C. "Can Reason Suppress the Fire Demon?" In Ken Sanders, Jack Durham, et al., eds., *Rangeland Fire Effects: A Symposium, November 27–29, 1984,* pp. 93–97. Boise: Bureau of Land Management and University of Idaho, 1984.

Leiberg, John B. *Forest Conditions in the Northern Sierra Nevada, California.* Department of Interior, U.S. Geological Survey, Professional Paper No. 8. Washington D.C.: Government Printing Office, 1902.

Leisz, Douglas R., and Carl C. Wilson. "To Burn or Not to Burn: Fire and Chaparral Management in Southern California." *Journal of Forestry* (February 1980): 94–95.

Leopold, Aldo. "Grass, Brush, Timber and Fire in Southern Arizona." *Journal of Forestry* 22 (1924): 1–10.

———. "'Piute Forestry' vs. Forest Fire Prevention." *Southwestern Magazine* 2 (1920): 12–13.

Leopold, A. Starker, S. A. Cain, C. H. Cottam, Ira N. Gabrielson, and T. L. Kimball. "The Leopold Committee Report: Wildlife Management in the National Parks." *American Forests* (April 1963): 32–35, 61–63.

———. "Let 'Em Burn." *Time*, October 28, 1974, p. 78.

Levine, Joel, ed. *Global Biomass Burning.* Cambridge, Mass.: MIT Press, 1991.

Lewis, Henry T. *Patterns of Indian Burning in California: Ecology and Ethno-history.* Ramona, Calif.: Ballena, 1973.

———. "Why Indians Burned: Specific Versus General Reasons." Paper presented at the Wilderness Fire Symposium, Missoula, Montana, November 15–18, 1983.

"Light Burning on Pine Forest." *American Forestry* 19, no. 10 (1913): 692.

Little, Charles. *The Dying of the Trees.* New York: Penguin, 1995.

———. "Smokey's Revenge." *American Forests* 99 (May–June 1993): 24–25, 58–60.

Lockmann, Ronald F. *Guarding the Forests of Southern California.* Glendale, Calif.: Clark, 1981.

Loeffelbein, Bob. "Smokey Hits 40!" *American Forests* (May 1984): 33.

Los Alamos National Laboratory. *A Special Edition of the SWEIS Yearbook Wildfire 2000.* LA-UR-00-3471. Los Alamos, N.M.: Los Alamos National Laboratory, August 2000.

Lyman, Chalmer K. "Our Choice: A Mild Singe or a Good Scorching." *Northwest Science* 21, no. 3 (1947): 129–133.

MacCleary, Doug. "Understanding the Role the Human Dimension Has Played in Shaping America's Forest and Grassland Landscapes." *EcoWatch* (February 10, 1994).

Machlis, G. E., A. B. Kaplan, S. P. Tuler, K. A. Bagby, and J. E. McKendry. *Burning Questions: A Social Science Research Plan for Federal Wildland Fire Management.* Report to the National Wildfire Coordinating Group. Boise: University of Idaho, 2002. Idaho Forest, Wildlife and Range Experiment Station, Contribution Number 943.

Maclean, John N. *Fire and Ashes: On the Front Lines Battling Wildfires.* New York: Holt, 2004.

———. *Fire on the Mountain: The True Story of the South Canyon Fire.* New York: Morrow, 1999.

Maclean, Norman. *Young Men & Fire.* Chicago: University of Chicago Press, 1992.

Manning, Richard. "Friendly Fire." *Sierra* (January–February 2001): 40–41, 110.

Manson, Marsden. "The Effect of the Partial Suppression of Annual Forest Fires in the Sierra Nevada Mountains." *Sierra Club Bulletin* 34 (January 1906): 22–24.

Maunder, Elwood R. "Voice from the South: Recollections of Four Foresters." Oral history interviews with Inman F. Eldredge, Walter J. Damtoft, Elwood L. Demmon, and Clinton H. Coulter. Forest History Society, Santa Cruz, California, 1977.

McBride, Joe R. "Managing National Parks." *Renewable Resources Journal* 11, no. 1 (Spring 1993): 24–25. In special report: "Workshop on National Parks Fire Policy: Goals, Perceptions, and Reality."

McCormick, W. C. "The Three Million." *American Forests* (August 1931): 479–480.

McDowell, John. "The Year They Firebombed the West." *American Forests* (May–June 1993): 22–23, 55.

McLaughlin, John S. "Restoring Fire to the Environment in Sequoia and Kings Canyon National Parks," 391–394. In *Proceedings: Tall Timbers Fire Ecology Conference, June 8–9, 1972.* Tallahassee, Fla.: Tall Timbers Research.

McManus, Reed. "Twice Burned? The Los Alamos Fire Rekindles Debate over Logging." *Sierra* (September–October 2000): 16–17.

McMlaran, M. P. "Comparison of Fire History Estimates between Open-scarred and Intact *Quercus Douglasii*." *American Midland Naturalist* 120 (1988): 432–435.

Miller, Char, ed. *American Forests: Nature, Culture, and Politics.* Lawrence: University Press of Kansas, 1997.

Minnich, Richard A. *The Biogeography of Fire in the San Bernardino Mountain of California.* Berkeley: University of California Press, 1988.

———. "Fire Mosaics in Southern California and Northern Baja California." *Science* 219 (March 18, 1983): 1287–1294.

———. "Landscapes, Land-use and Fire Policy: Where Do Large Fires Come From?" In J. M. Moreno, ed., *Large Forest Fires,* pp. 133–158. Leiden, Netherlands: Backhuys, 1998.

Minnich, Richard A., and Yue Hong Chou. "Wildland Fire Patch Dynamics in the Chaparral of Southern California and Northern Baja California." *International Journal of Wildland Fire* 7, no. 3 (1997): 221–248.

Monastersky, R. "Burning Questions." *Science News* 138 (1990): 264–266.

Moore, William R. "From Fire Control to Fire Management." *Western Wildlands* 1, no. 3 (1974): 11–15.

Morrison, Ellen Earnhardt. *Guardian of the Forest: A History of the Smokey Bear Program.* New York: Vantage, 1976.

———. *The Smokey Bear Story.* Alexandria, Va.: Morielle, 1995.

Morrison, Micah. *Fire in Paradise: The Yellowstone Fires and Politics of Environmentalism.* New York: HarperCollins, 1993.

Muir, John. *The Mountains of California.* 1894. Reprint. New York: American Museum of Natural History and Doubleday, 1961.

Murkowski, Frank H. "Lessons Must Be Learned from Los Alamos Fire." Press Release from Senator Frank H. Murkowski, Alaska, chair of Senate Committee on Energy and Natural Resources, July 27, 2000. Available at http://energy.senate.gov/press/releases/losalamos.firehearing.htm.

Murphy, E. W. "California Pays the Red Piper." *American Forests* (April 1939): 202. Murphy, James L., and Frank T. Cole. "Villains to Heroes: Overcoming the Prescribed Burner versus Forest Firefighter Paradox." In *Proceedings: 20th Tall Timbers Fire Ecology Conference,* pp. 17–22. Tallahassee, Fla.: Tall Timbers Research, 1998.

Murphy, R. W. "Experimental Burning in Park Management." In *Proceedings from the California Tall Timbers Fire Ecology Conference, November 9–10, 1967*, pp. 207–216.

Mutch, Robert W. "Fighting Fire with Prescribed Fire: A Return to Ecosystem Health." *Journal of Forestry* (November 1994): 31–33.

———. "I Thought Forest Fires Were Black!" *Western Wildlands* 1, no. 3 (1974): 16–21.

———. "Understanding Fire as Process and Tool." Adapted from "Fire Management Today: Tradition and Change in the Forest Service." Paper presented at Society of American Foresters National Convention, Washington, D.C., September 28–October 2, 1975.

———. "Wilderness Fires Allowed to Burn More Naturally." *Fire Control Notes* 33 (1972): 3–6.

———. "Wildland Fires and Ecosystems: A Hypothesis." *Ecology* 51 (1970): 1046–1051.

———. "Will We Be Better Prepared for the Fires of 2006?" *Bugle: Journal of the Rocky Mountain Elk Foundation* (March–April 2001).

National Research Institute. *How Can We Live with Wildland Fire?* Davis: California Communities Program, University of California, 1995.

Nelson, Robert H. *A Burning Issue: A Case for Abolishing the U.S. Forest Service*. New York: Rowman and Littlefield, 2000.

Oberle, Mark. "Forest Fires: Suppression Policy Has Its Ecological Drawbacks." *Science* 165 (August 1969): 568–571.

Oettmeier, W. M. "The Place of Prescribed Burning." *Forest Farmer* (May 1956): 6, 7, 18, 19.

Olmsted, Frederick E. "Fire and the Forest: The Theory of 'Light Burning.'" *Sierra Club Bulletin* 8 (January 1911): 42–47.

———. "Forest Devastation: A National Danger and a Plan to Meet It." *Journal of Forestry* 17 (1919): 911–935.

Ong, D. "Fighting Fire with Fire: How Fire Plays a Beneficial Role in the Health of the Forest." *UC Davis Magazine* (May–June 1989): 8–11.

Ostrander, H. J. "How to Save the Forests by Use of Fire." *San Francisco Call*. Letter to Editor, September 23, 1902, p. 6.

Parsons, D., L. Bancroft, T. Nichols, and T. Stohlgren. "Information Needs for Natural Fire Management Planning." In J. F. Lotan, B. M. Kilgore, W. C. Fischer, and R. F. Mutch, eds., *Proceedings: Symposium and Workshop on Wilderness Fire, 15–18 November 1983, Missoula, Montana*. USDA Forest Service General Technical Report INT-182.

Parsons, D. J., and S. H. DeBenedetti. "Impact of Fire Suppression on a Mixed-Conifer Forest." *Forest Ecology* 2 (1979): 21–33.

Parsons, David J. "Objects or Ecosystems? Giant Sequoia Management in National Parks." In *Proceedings of the Symposium on Giant Sequoias: Their Place in the Ecosystem and Society, 23–25 June 1992, Visalia, California*. USDA Forest Service General Technical Report PSW-GTR-151.

———. "Prescribed Fire Review Sparks Studies of Giant Sequoia–Fire Interactions." *Park Science* 9 (1989): 19.

———. "The Role of Fire Management in Maintaining Natural Ecosystems." In *Fire Regimes and Ecosystem Properties, 1981.* USDA Forest Service General Technical Report GTR-WO-26.

———. "The Role of Fire in Natural Communities: An Example from the Southern Sierra Nevada." *Environmental Conservation* 3, no. 2 (1976): 91–99.

———. "The Role of Fire in Park Management." *Parks* 2 (1977): 1–4.

Parsons, David J., and Stephen J. Botti. "Restoration of Fire in National Parks." In *The Use of Fire in Forest Restoration.* General Technical Report INT-GTR-341. June 1996. Available at http://www.fs.fed.us/rm/pub/int_gtr341/gtr341_4.html.

Parsons, David J., and Jan W. van Wagtendonk. "Fire Research and Management in the Sierra Nevada National Parks." In W. L. Halvorson and G. E. Davis, eds., *Ecosystem Management in the National Parks,* pp. 25–48. Tucson: University of Arizona Press, 1996.

Parsons, D. S., D. M. Graber, J. K. Agee, and J. van Wagtendonk. "Natural Fire Management in National Parks." *Environmental Management* 10 (1986): 21–24.

Phillips, J. "The Crisis in Our Forests." *Sunset Magazine* (July 1995): 87–92.

Pinchot, Gifford. *Breaking New Ground.* 1947. Reprint. Washington, D.C.: Island, 1998.

———. *The Fight for Conservation.* New York: Doubleday, Page, 1910.

———."The Relation of Forests and Forest Fires." *National Geographic* 10 (1899): 393–403.

Plumb, T. R. "Response of Oaks to Fire." In *Proceedings of the Symposium on the Ecology, Management, and Utilization of California Oaks, June 26–28, 1979, Claremont, California.* USDA Forest Service General Technical Report PSW-44.

Plummer, Fred G. *Forest Fires: Their Causes, Extent and Effects, with a Summary of Recorded Destruction and Loss.* USDA Forest Service Bulletin 117. Washington, D.C.: Government Printing Office, 1912.

Pyne, Stephen J. *America's Fires: Management of Wildlands and Forests.* Durham, N.C.: Forest History Society, 1997.

———. *Fire in America: A Cultural History of Wildland and Rural Fire.* 1982. Reprint. Seattle: University of Washington Press, 1997.

———. "Firestick History." *Journal of American History* 76, no. 4 (1990): 1132–1141.

———. "The Fires This Time, and Next." *Science* 294 (November 2001): 1005–1006.

———. *Smokechasing.* Tucson: University of Arizona Press, 2003.

———. *World Fire: The Culture of Fire on Earth.* New York: Holt, 1995.

———. *Year of the Fires.* New York: Viking Penguin, 2001.

Pyne, Stephen, Patricia Andrews, and Richard Laven. *Introduction to Wildland Fire,* 2d ed. New York: Wiley, 1996.

Rasmussen, Matt. "The Long Reach of Humanity." *Forest Magazine* (March–April 2000): 14–19.

Reddington, Paul G. "What Is the Truth? Conclusion of the Light-burning Controversy." *Sunset: The Pacific Monthly* 44 (June 1920): 56–58.

Rice, Carol. "A Balanced Approach: Dr. Biswell's Solution to Fire Issues in Urban Interface and Wildland Ecosystems." USDA Forest Service General Technical Report

PSW-GTR-158. First presented at the Biswell Symposium: Fire Issues and Solutions in Urban Interface and Wildland Ecosystems, February 15–17, 1994. Walnut Creek, California, 1995.

Rodgers, Andrew Denny, III. *Bernard Edward Fernow: A Story of North American Forestry.* Princeton, N.J.: Princeton University Press, 1951.

Romme, W. H. "Fire and Landscape Diversity in Subalpine Forests of Yellowstone National Park." *Ecological Monographs* 52 (1982): 199–221.

Romme, W. H., and D. G. Despain. "Historical Perspective on the Yellowstone Fires of 1988." *BioScience* 32 (1982): 695–699.

Romme, W. H., and D. H. Knight. "Landscape Diversity: The Concept Applied to Yellowstone Park." *BioScience* 32 (1982): 664–670.

Romme, W. H., M. G. Turner, R. H. Gardner, W. W. Hargrove, G. A. Tuskan, D. G. Despain, and R. A. Renkin. "A Rare Episode of Sexual Reproduction in Aspen (*Populus Tremuloides Michx*) following the 1988 Yellowstone Fires." *Natural Areas Journal* 17 (1997): 17–25.

Romme, W. H., L. L. Wallace, and J. S. Walker. "Aspen, Elk, and Fire in Northern Yellowstone National Park." *Ecology* 76, no. 7 (1995): 2097–2106.

Rothman, Hal K., ed. *"I'll Never Fight Fire with My Bare Hands Again": Recollections of the First Forest Rangers of the Inland Northwest.* Lawrence: University Press of Kansas, 1994.

Rundel, P. W. "The Relationship between Basal Fire Scars and Crown Damage in Giant Sequoia." *Ecology* 54 (1973): 210–213.

Rundel, P. W., and D. J. Parsons. "Structural Changes along a Fire-Induced Age Gradient." *Journal of Range Management* 32, no. 6 (1979): 462–466.

Runte, Alfred. *National Parks: The American Experience*, 2d ed. Lincoln: University of Nebraska Press, 1987.

Sando, R. W. "Natural Fire Regimes and Fire Management: Foundations for Direction." *Western Wildlands* 4, no. 4 (1978): 34–44.

Schiff, Ashley L. *Fire and Water: Scientific Heresy in the Forest Service.* Cambridge, Mass.: Harvard University Press, 1962.

Schimke, Harry E., and Lisa R. Green. "Prescribed Fire for Maintaining Fuel-Breaks in the Central Sierra Nevada." Berkeley, Calif.: Pacific Southwest Forest and Range Experiment Station, 1970.

Schultz, A. M., and H. H. Biswell. "Reduction of Wildfire Hazard." *California Agriculture* 10, no.11 (1956): 4–5.

Sellars, Richard West. *Preserving Nature in the National Parks.* New Haven, Conn.: Yale University Press, 1997.

Shankland, Robert. *Steve Mather of the National Parks.* New York: Knopf, 1951.

Shea, John P. "Our Pappies Burned the Woods." *American Forests* (April 1940): 159–174.

Shipek, Florence. "Kumeyaay Plant Husbandry: Fire, Water, and Erosion Management Systems." In Thomas C. Blackburn and Kat Anderson, eds., *Before the Wilderness: Environmental Management by Native Californians.* Menlo Park, Calif.: Ballena, 1993.

Shoemaker, Len. *Saga of a Forest Ranger: A Biography of William R. Kreutzer, Forest Ranger No. 1.* Boulder: University of Colorado Press, 1958.

Show, Stuart Bevier. "National Forests in California: An Interview Conducted by Amelia Roberts Fry." Berkeley: University of California, Regional Cultural History Project, 1965.

———. "Personal Reminiscences of a Forester 1907–1931." Written at the request of R. E. McArdle, chief, U.S. Forest Services. Privately published. Berkeley, California, 1955.

Show, S. B., and E. I. Kotok. "Fire and the Forest (California Pine Region)." USDA, Department Circular 358. Washington, D.C., August 1925.

———. "Forest Fires in California, 1991–1920: An Analytical Study." USDA, Department Circular 243. Washington, D.C., February 1923.

———. "The Role of Fire in the California Pine Forest." USDA, Department Bulletin No. 1294. Washington, D.C., December 1924.

Smith, Conrad. *Media and Apocalypse: News Coverage of the Yellowstone Forest Fires, the* Exxon Valdez *Oil Spill, and the Loma Prieta Earthquake.* Westport, Conn.: Greenwood, 1992.

Society of American Foresters. "Fire as a Tool in Forest Protection and Management." *Proceedings: Society of American Foresters, Southern California Section, Annual Meeting, December 1, 1962, Oakland, California* (Society of American Foresters: Bethesda, Md., 1962).

Steen, Harold K. *The U.S. Forest Service: A History.* Seattle: University of Washington Press, 1976.

Stephenson, N. L., D. J. Parsons, and H. T. Nichols. "Replies from the Fire Gods." *American Forests* 96 (1990): 35, 70.

Sterling, E. A. "Attitude of Lumbermen toward Forest Fires." In *Yearbook of the United States Department of Agriculture, 1904*, pp. 133–140. Washington, D.C.: Government Printing Office, 1905.

Stoddard, H. L. "The Use of Fire in Pine Forests and Game Lands of the Deep Southeast." In *Proceedings: First Tall Timbers Fire Ecology Conference*, pp. 31–42. Tallahassee, Fla.: Tall Timbers Research Institute, 1962.

Stolzenburg, W. "Fire in the Rain Forest." *Nature Conservancy Magazine* (May–June 2001): 22–27.

———. "Reapers of the Flame." *Nature Conservancy Magazine* (May–June 2001): 10–11.

Strohmaier, David J. *The Seasons of Fire: Reflections on Fire in the West.* Reno: University of Nevada Press, 2001.

Suckling, Kieran. "Fire & Forest Ecosystem Health in the American Southwest." *Southwest Forest Alliance and Southwest Center for Biological Diversity*, 1996. Available at http://www.sw-center.org/swcbd/papers/fire-prm.html.

Sullivan, Margaret. *Firestorm! The Story of the 1991 East Bay Fire in Berkeley.* Berkeley, Calif.: City of Berkeley, 1993.

Sweeney, J. R., and H. H. Biswell. "Quantitative Studies of the Removal of Litter and Duff by Fire under Controlled Conditions." *Ecology* 42, no. 3 (1961): 572–575.

Swetnam, T. W. "Fire History and Climate Change in Giant Sequoia Groves." *Science* 262 (1993): 885–889.

Task Force on California's Wildland Fire Problem. *Recommendations to Solve California's Wildland Fire Problem.* California Department of Conservation, June 1972.

Taylor, Ron. "Fire in the Redwoods." *Westways* (August 1968): 36–37.

Thurmond, Jack. "Through 1930 with the Dixie Crusaders." *American Forests* (March 1930): 151.

Timbrook, Jan, John R. Johnson, and David D. Earle. "Vegetation Burning by the Chumash." In Thomas C. Blackburn and Kat Anderson, eds., *Before the Wilderness: Environment Management by Native Californians.* Menlo Park, Calif.: Ballena, 1993.

Turner, M. G., W. H. Romme, and D. B. Tinker. "Surprises and Lessons from the 1988 Yellowstone Fires." *Frontiers in Ecology and the Environment* 1 (2003): 351–358.

Tweed, W. "Born of Fire." *National Parks Magazine* 61 (1987): 23–27, 45.

U.S. Bureau of Land Management. "Burning Issues: An Interactive Multimedia Program." Joint project of the Bureau of Land Management and Florida State University. Tallahassee: Florida State University, 2000.

———. *Using Fire to Manage Public Lands.* Boise, Idaho: BLM National Office of Fire and Aviation, 1997. Pamphlet.

U.S. Department of Agriculture, Forest Service. *Living with Fire.* Southwestern Region, 2000.

———. *Rx Fire!* Southwestern Region, 1994. Pamphlet.

———. *The True Story of Smokey Bear.* Western Publishing, 1960. Comic book.

U.S. Department of the Interior. *Review and Update of the 1995 Federal Wildland Fire Management Policy: Report to the Secretaries of the Interior, of Agriculture, of Energy, of Defense, and of Commerce; the Administrator, Environmental Protection Agency; the Director, Federal Emergency Management Agency; and the National Association of State Foresters, by an Interagency Federal Wildland Fire Policy Review Working Group.* Boise, Idaho: National Interagency Fire Center, January 2001.

U.S. General Accounting Office. "Western National Forests: A Cohesive Strategy Is Needed to Address Catastrophic Wildfire Threats." GAO-RCED-99-65. Washington, D.C., April 2, 1999.

U.S. National Park Service. "Cerro Grande Prescribed Fire Board of Inquiry Final Report." National Park Service, February 26, 2001. Available at http://www.nps.gov/fire/fireinfo/cerrogrande/reports/Board_report-feb26final.pdf.

———. "Cerro Grande Prescribed Fire Investigation Report." May 18, 2000. Available at http://www.nps.gov/cerrogrande.htm.

———. "The Yellowstone Fires: A Primer on the 1988 Fire Season." Yellowstone National Park, October 1, 1988.

Vale, Thomas. *Fire, Native Peoples, and the Natural Landscape.* Washington, D.C.: Island, 2002.

———. "The Myth of the Humanized Landscape: An Example from Yosemite National Park." *Natural Areas Journal* 18, no. 3 (1999): 231–236.

Vankat, J. L. "Fire and Man in Sequoia National Park." *Annuals of the Association of American Geographers* 67 (1977): 17–27.

van Wagtendonk, Jan W. "Dr. Biswell's Influence on the Development of Prescribed Burning in California." In USDA Forest Service General Technical Report PSW-

GTR-158, *The Biswell Symposium: Fire Issues and Solutions in Urban and Wildland Ecosystems, February 15–17, 1994*, pp. 11–15.

———. "Fire Management in the Yosemite Mixed-Conifer Ecosystem." In J. F. Lotan, B. M. Kilgore, W. C. Fischer, and R. F. Mutch, eds., *Proceedings: Symposium and Workshop on Wilderness Fire, 15–18 November 1983, Missoula, Montana.* USDA Forest Service General Technical Report INT-182.

———. "Fire Suppression Effects on Fuels and Succession in Short-fire Interval Wilderness Ecosystems." In J. F. Lotan, B. M. Kilgore, W. C. Fischer, and R. F. Mutch, eds., *Proceedings: Symposium and Workshop on Wilderness Fire, 15–18 November 1983, Missoula, Montana.* USDA Forest Service General Technical Report INT-182.

———. "GIS Applications in Fire Management Research." In S. C. Nodvin and T. A. Waldrop, eds., *Fire and the Environment: Ecological and Cultural Perspectives: Proceedings of an International Symposium 1991.* USDA Forest Service Southeastern Forest Experimental Station General Technical Report GTR-SE-69.

———. "Large Fires in Wilderness Areas." In J. K. Brown, R. W. Mutch, C. W. Spoon, R. H. Wakimoto, eds., *Proceedings: Symposium on Fire in Wilderness and Park Management, 30 March–1 April 1993, Missoula, Montana.* USDA Forest Service General Technical Report INT-GTR-320.

———. "Park Goals and Current Fire Policy." *Renewable Resources Journal* 11, no. 1 (Spring 1993): 19. In special report: "Workshop on National Parks Fire Policy: Goals, Perceptions, and Reality."

———. "The Role of Fire in the Yosemite Wilderness." Paper presented at the National Wilderness Research Conference, Fort Collins, Colorado, July 23–26, 1985.

———. "Spatial Patterns of Lightning Strikes and Fires in Yosemite National Park." In *Proceedings of the 12th Conference on Fire and Forest Meteorology, 26–28 October 1993, Jekyll Island, GA.* Bethesda, Md.: Society of American Foresters.

———. "Wilderness Fire Management in Yosemite National Park." In E. A. Schofield, ed., *Earthcare: Global Protection of Natural Areas: Proceedings of the Fourteenth Biennial Wilderness Conference.* Boulder, Colo.: Westview, 1978.

Vitas, George. *Forest and Flame in the Bible.* A Program Aid of the Cooperative Forest Fire Prevention Campaign Sponsored by the Advertising Council, State Foresters, and the U.S. Department of Agriculture, Forest Service. PA-93. Reprinted December 1961.

Vogl, Richard J. "Comments on Controlled Burning." In *Proceedings: Tall Timbers Fire Ecology Conference, April 10–11, 1969*, pp. 1–4. Tallahassee, Fla.: Tall Timbers Research Station, 1969.

———. "Smokey's Mid-Career Crisis." *Saturday Review of the Sciences* 1, no. 2 (March 1973): 23–29.

Wagle, R. F., and Thomas W. Eakle. "A Controlled Burn Reduces the Impact of a Subsequent Wildfire in a Ponderosa Pine Vegetation Type." *Forest Science* 25, no. 1 (1979): 123–128.

Wagner, W. W. "Past Fire Incidence in Sierra Nevada Forests." *Journal of Forestry* 59 (1961): 739–748.

Wahlenberg, W. G., S. W. Greene, and H. R. Reed. "Effects of Fire and Cattle Grazing on Longleaf Pine Lands, as Studied at McNeill, Miss." Technical Bulletin No. 683. Washington, D.C.: U.S. Department of Agriculture, June 1939.

Wakimoto, R. H. "The Yellowstone Fire of 1988: Natural Processes and National Policy." *Northwest Science* 64 (1990): 239–242.

Walker, T. B. "T. B. Walker Expresses His Views on Conservation." *San Francisco Chronicle*, January 5, 1913, p. 56.

Weaver, Harold. "Effects of Prescribed Burning in Ponderosa Pine." *Journal of Forestry* 55 (February 1957): 823–826.

———. "Fire as an Ecological Factor in the Southwestern Ponderosa Pine Forests." *Journal of Forestry* 49 (February 1951): 93–98.

———. "Fire as an Ecological and Silvicultural Factor in the Ponderosa-Pine Region of the Pacific Slope." *Journal of Forestry* 41 (January 1943): 7–15.

———. "Fire as an Enemy, Friend, and Tool in Forest Management." *Journal of Forestry* 53 (July 1955): 499–504.

———. "Fire and Management Problems in Ponderosa Pine." In *Proceedings: 3rd Annual Tall Timbers Fire Ecology Conference, April 9–10, 1964*, pp. 61–79. Tallahassee, Fla.: Tall Timbers Research Station, 1964.

———. "Implications of the Klamath Fires of September 1959." *Journal of Forestry 59* (August 1961): 569–572.

Weaver, Harold, and Harold Biswell. "How Fire Helps the Big Trees." *National Parks & Conservation Magazine* 43 (1969): 16–19.

Weldon, Leslie A. C. "The Use of Fire in Forest Restoration: Dealing with Public Concerns in Restoring Fire to the Forest." General Technical Report INT-GTR-341. USDA Forest Service, Rocky Mountain Research Station. Available at http://www.fs.fed.us/rm/pubs/int_ugtr341/gtr341_6.html.

Wheeler, H. N. "Controlled Burning in Southern Pine." *Journal of Forestry* 42 (June 1944): 449.

Whelan, Robert J. *The Ecology of Fire*. Cambridge, Mass.: Cambridge University Press, 1995.

White, John R. "Letters to the *Times*: Scare Heads Mislead." *Los Angeles Times*, August 29, 1928. Letter to editor from superintendent, Sequoia National Park.

White, John R., and W. Fry. *Big Trees*. Palo Alto, Calif.: Stanford University Press, 1931.

White, Stewart E. "Getting at the Truth: Is the Forest Service Really Trying to Lay Bare the Facts of the Light-Burning Theory?" *Sunset: The Pacific Monthly* (May 1920): 62, 80–82.

———. "Woodsmen, Spare Those Trees! Our Forests Are Threatened: A Plea for Protection." *Sunset: The Pacific Monthly* (March 1920): 23–26, 108–117.

Wilkinson, T. "Prometheus Unbound." *Nature Conservancy Magazine* (May–June 2001): 12–20.

Williams, Ted. "Burning Money." *Audubon* 103 (January–February 2001): 34–41.

———. "Incineration of Yellowstone." *Audubon* (January 1989): 38–85.

———. "Only You Can Postpone Forest Fires." *Sierra* 80 (1995): 36–43.

Wilson, Carl C., and James B. Davis. "Forest Fire Laboratory at Riverside and Fire Research in California: Past, Present, and Future." General Technical Report PSW-105. Berkeley, Calif.: Pacific Southwest Forest and Range Experiment Station, May 1988.

Wooley, H. E. "What Has Been Accomplished in Fire Protection on the National Forests." *American Forestry* 19 (November 1913).

Wright, H. A., and A. W. Bailey. *Fire Ecology: United States and Canada.* New York: Wiley Interscience, 1982.

Wuerther, G. "Fire Power." *National Parks & Conservation Magazine* (May–June 1995): 32–37.

Zinke, P. J., and R. L. Crocker. "The Influence of Giant Sequoia on Soil Properties." *Forest Science* 8 (1962): 2–11.

Index